ENVIRONMENTAL RESEARCH ADVANCES

PLANT INVASION ECOLOGY

IMPACTS AND SUSTAINABLE MANAGEMENT

ENVIRONMENTAL RESEARCH ADVANCES

Additional books in this series can be found on Nova's website under the Series tab.

Additional e-books in this series can be found on Nova's website under the e-book tab.

ENVIRONMENTAL RESEARCH ADVANCES

PLANT INVASION ECOLOGY

IMPACTS AND SUSTAINABLE MANAGEMENT

PRABHAT KUMAR RAI, PH.D.

New York

Copyright © 2013 by Nova Science Publishers, Inc.

For permission to use material from this book please contact us:
Telephone 631-231-7269; Fax 631-231-8175
Web Site: http://www.novapublishers.com

NOTICE TO THE READER

The Publisher has taken reasonable care in the preparation of this book, but makes no expressed or implied warranty of any kind and assumes no responsibility for any errors or omissions. No liability is assumed for incidental or consequential damages in connection with or arising out of information contained in this book. The Publisher shall not be liable for any special, consequential, or exemplary damages resulting, in whole or in part, from the readers' use of, or reliance upon, this material. Any parts of this book based on government reports are so indicated and copyright is claimed for those parts to the extent applicable to compilations of such works.

Independent verification should be sought for any data, advice or recommendations contained in this book. In addition, no responsibility is assumed by the publisher for any injury and/or damage to persons or property arising from any methods, products, instructions, ideas or otherwise contained in this publication.

This publication is designed to provide accurate and authoritative information with regard to the subject matter covered herein. It is sold with the clear understanding that the Publisher is not engaged in rendering legal or any other professional services. If legal or any other expert assistance is required, the services of a competent person should be sought. FROM A DECLARATION OF PARTICIPANTS JOINTLY ADOPTED BY A COMMITTEE OF THE AMERICAN BAR ASSOCIATION AND A COMMITTEE OF PUBLISHERS.

Additional color graphics may be available in the e-book version of this book.

Library of Congress Cataloging-in-Publication Data

Rai, Prabhat Kumar.
  Plant invasion ecology : impacts and sustainable management / Prabhat Kumar Rai.
     pages cm
  Includes bibliographical references and index.
  ISBN: 978-1-62948-111-1 (hardcover)
  1. Plant invasions. 2. Invasive plants--Ecology. 3. Invasive plants--Control. I. Title.
  SB613.5.R35 2013
  333.95'33--dc23
                          2013034671

Published by Nova Science Publishers, Inc. † *New York*

*In the loving memory of my grand parents-
late Shri Braj Bihari Rai and Shrimati Chandravati rai*

Contents

Preface

Exotic invasive plant species poses serious threat to the native biodiversity. Invasive plants transmogrify the landscape ecology in a highly complex manner leading to a sort of ecological explosion. Global terrestrial as well as aquatic ecosystems are invaded by various invasive plant species. Invasive species are alien species whose introduction and spread threatens ecosystems, habitats or species with socio-cultural, economic and/or environmental harm, and harm to human health. Present book aims to provide a critical review on mechanisms, impact and management of invasive species particularly in context of plants. Plant invasion is now increasingly being recognised as global problem and various continents are adversely affected, although to a differential scale. Invasive plants not only alter plant ecosystem function but also result in large economic costs from lost ecosystem services. Quest for the ecological mechanism lying behind the success of invasive species over native species has drawn the attention of researches worldwide particularly in context of diversity-stability relationship. Transport, colonization, establishment and landscape spread may be different steps in success of invasive plants and each and every step is checked through several ecological attributes. Further, several ecological attribute and hypothesis (enemy release, novel weapon, empty niche, evolution of increased competitive ability etc.) were proposed pertaining to success of invasive plant species. However, single theory will not be able to account for invasion success among all environments as it may vary spatially and temporally. Therefore, in order to formulate a sustainable management plan for invasive plants, it is necessary to develop a synoptic view of the dynamic processes involved in the invasion process. Moreover, invasive species can act synergistically with other elements of global change, including land-use change, climate change, increased concentrations of atmospheric carbon dioxide and nitrogen deposition. Henceforth, a unified framework for biological invasions that reconciles and integrates the key features of the most commonly used invasion frame-works into a single conceptual model that can be applied to all human-mediated invasions.

Acknowledgment

I consider it a rare opportunity, to thank Professor A.N. Rai, Director NAAC and Former Vice Chancellor of Mizoram University as well as North Eastern Hill University-NEHU) for his blessings, encouragement and support.

I am thankful to Professor Lalthantluanga, Vice Chancellor, Mizoram University for the all sort of support. I am extremely thankful to Professor R.P. Tiwary, Department of Geology, Mizoram University who always helped and encouraged me. I am thankful to Professor B. Gopichand, Dean, School of ES& NRM and Head Department of Environmental Science, Mizoram University. Pertaining to academic guidance I am thankful to Professor Diwakar Tiwari, Professor B.P. Nautiyal and Professor V.P. Khanduri.

I am thankful to my research team namely Lalita Panda, Biku, Pallab Deb, and Muni Singh for their support. I would like to thank my friends like Mr. Alok Chourasia, K.S. Kumar, Sanju Sonkar, Atul Sharma, Ramchandra, Nimesh Rai and Mukesh Rai for always being with me.

I would like to thank my parents (Father-Dr. Om Prakash Rai, Mother-Mrs. Usha Rai), family (Brother-Prashant Rai, Sister-Pratibha Rai) and my wife (Garima Rai) who have supported and encouraged me all through this course. Further, I am thankful to my brother in law Dr. Ved Prakash Rai of Genetics and Plant Breeding, BHU for all his affection and encouragement. Moreover, I would like to extend my love to little Pranjali and Rachit as they brought a great deal of fortune with them.

Prabhat Kumar Rai
Senior Assistant Professor, Department of Environmental Sciences
School of Earth Sciences and Natural Resource Management
Mizoram Central University, Tanhril, Aizawl, 796004, India
Email: prabhatrai24@gmail.com

About the Author

Dr. Prabhat Kumar Rai is currently Senior Assistant Professor in Department of Environmental Science, School of Earth Science and Natural Resource Management, Mizoram University, Aizawl, India. Dr. Rai did his whole education i.e., from schooling (from Central Hindu School) to Ph.D. from Banaras Hindu University (B.H.U.), Varanasi, India. He is recipient of many national fellowships/awards i.e., CSIR-JRF, SRF, GATE, D.B.T. Overseas award, and DST young scientist awards. Dr. Rai received third best research presentation award from American Academy of Sciences. Dr. Rai published fifty research papers (Including ca ten Review papers) in leading journals of Elsevier, Springer and Taylor & Francis. Also, Dr. Rai attended thirty five international and national conferences/seminar/workshop/symposia. Pertaining to his research, Dr. Rai visited American Academy of Sciences, Houston, United States of America, and Chinese Academy of Sciences at Nanjing, China and University of Hasselt, Brussels, Belgium. His book entitled *Heavy Metal Pollution and its Phytoremediation though wetland Plants* is already published with Nova Science Publishers, New York, USA.

Chapter 1

Introduction to Invasion Ecology

Biodiversity is an integral component of environment responsible for various life processes and hence persistence of life on this planet. Biodiversity is inextricably linked with sustainable development (Rai, 2011; Rai, 2012; Rai, 2013; Rai & Rai, 2013). Biodiversity is extremely precious resource issue of 21st century as it is intimately linked to economy as well as environment (Rai, 2009; Rai & Lalramnghinglova, 2010 a, b, c; Rai & Lalramnghinglova, 2011a, b; Rai, 2012). In recent era of rapid industrialization and urbanization, anthropogenic perturbations are causing a biodiversity crisis, with species extinction rates up to 1000 times higher than background (Brooks et al., 2006). Plant invasions have caused an unprecedented loss of the global native biodiversity. Henceforth, in the current era of Anthropocene, unravelling the facts that what makes the replacement of indigenous climax communities originating through succession by exotic invasive species (Lovel, 1997; Rejmánek, 1999; McNeely, 2001; Van Kleunen et al., 2010; Crossman et al., 2011) is of extreme importance.

Plant invasion is one of the major threats to global biodiversity, being 'big five' environmental issues of public concern (Sala et al., 2000; Didham et al., 2005) and one of the six most serious environmental problems which may influence future economical and social development (Gewin, 2005; Mooney et al., 2005; Huang et al., 2008). Earth Summit in Rio de Janeiro, 1992 regarded invasive species as one of the main reasons for the loss of biodiversity (Keane & Crawley, 2002; Born et al., 2005). Through advent of communication, science and technology, various factors like transport, migration, and commerce, humans are continuing to disperse an ever-increasing array of species across previously insurmountable environmental barriers such as oceans, mountain ranges, rivers, and inhospitable climate zones (Mack et al., 2000).

Humans have extensively altered the global environment, changing global biogeochemical cycles, transforming land and enhancing the mobility of biota (Chapin et al., 2000). Further, high natural resource extraction, short food chains, food web simplification, habitat homogeneity, landscape homogeneity, heavy use of herbicides, pesticides, and insecticides, large importation of non-solar energy, large importation of nutrient supplements, convergent soil characteristics, modified hydrological cycles, reduced biotic and physical disturbance regimes, global mobility of people, goods, and services are some characteristics of intentionally modified ecosystems (Western, 2001). Gallagher and Carpenter (1997) remarked on human-dominated ecosystems, the concept of a pristine ecosystem, untouched by human activity, "is collapsing in the wake of scientists' realization that there are *no places left on Earth that doesn't fall under humanity's shadow.*"

Initially trade and then transportation systems of various kinds have distributed invasive species around the world (Barlow, 1997; Jabolonski, 1991); as Kaiser (1999) notes, "The world's ecosystems will never revert to the pristine state they enjoyed before humans began to routinely crisscross the globe." Even those considered "natural" almost inevitably contain invasive species, frequently in dominant roles. Moreover, modern technological systems continue to increase the scale of these impacts (Palumbi, 2001). Recent advances in genetics and molecular biology also paved the way for an open debate on their impacts on ecology and global biodiversity. The histories of invasions and agriculture are intimately linked, with many crop and livestock pests being invasive species and vice versa (Guillemaud et al., 2011). Particularly, agricultural biotechnology, i.e., the insertion of genes into crops, has generated concern over the risk of producing new invasive species or exacerbating current weed problems (Parker & Kareiva, 1996; Guillemaud et al., 2011). Also, it has been demonstrated that the sale and transport of prohibited invasive plants and their misidentification present the greatest risk associated with the horticultural trade (Maki & Galatowitsch, 2004).

In the light of aforesaid, it is clear that modern intensive agriculture paved the way for invasive plants through trade and transport all across the globe. Further, it leads to land use changes (conversion of forests/grasslands into agroecosystems), habitat fragmentation as well as increase the level of persistent organic pollutants (POPs), hence, resulting in the increased level of CO_2/climate change. All these factors are directly or indirectly linked to biological invasions, in totality resulting in biodiversity loss (Figure 1).

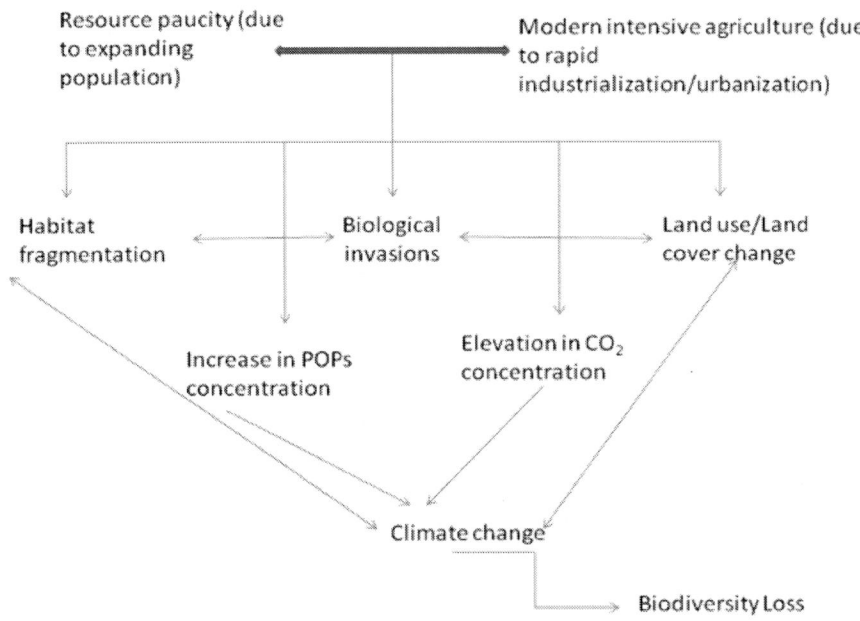

Figure 1. Paradigm of global interlinked ecological issues or concerns [Modified after Vitousek et al. (1996, 1997); Didham et al. 2000].

Invasion ecology is not very old field and attained pace from 20th century, and mostly the researches in its multifaceted disciplines attained pace during last fifty years (Richardson, 2011). Although the first paper on species invasions (Grinnel, 1919) appeared in 1919, study of the phenomenon is often traced back to Darwin's *Beagle* voyage, when he documented many European plants thriving as aliens in South America. He pointed out that escape from the parasites and diseases that attack them in their native range may contribute to the rapid spread of invading plants and animals (Sax et al., 2005).

In eighteenth century, plant invasion was generally being recognized as ecological phenomenon. However, in current scenario, invasive species are the second largest threat to global biodiversity just after habitat destruction and is the prime factor responsible for species extinction in most island states (Schei, 1996; Sharma et al., 2005). Global terrestrial as well as aquatic ecosystems are invaded by various invasive plant species (Usher et al., 1988; Lonsdale 1999; Totland et al., 2005). In present era of ecological sciences, concept of plant invasion may be treated as a sort of ecological explosion (Elton, 1958). Invasion of exotic invasive species is among the most important ecological perturbation experienced by natural ecosystems (Sharma et al., 2005). Such perturbation drives and regulates the dynamic ecosystems (Sousa, 1984; Gurevitch & Padilla, 2004; Sharma et al., 2005). In order to understand plant community patterns, it has been suggested to investigate that plant-plant replacements, first by sampling long-term vegetation plots in order to map them, and then by manipulating mechanisms and tolerances in field experiments in order to understand what causes them (Myster, 2012).

Several reviews have enriched the ecological literatures with the several facets of invasion ecology, however, multifaceted issues in totality, associated with this ecological phenomenon were lacking. Present book aims to provide a critical review on mechanisms, impact and management of invasive species particularly in context of plants. However, at some places, biological invasions would be described in terms of animals also. Now, before going into details of the present review, it is pertinent to understand the generalized definitions and terms associated with invasion ecology. Davis (2009) in his book mentioned different views and terminology imposed in invasion ecology and unlikely to reach any consensus.

Definition

The Convention on Biological Diversity held in 1992, defines an invasive species as "an alien species whose introduction and spread threatens ecosystems, habitats or species with socio-cultural, economic and/or environmental harm, and/or harm to human health" (COP, 2002; Born et al., 2005; Davis, 2009).

Whereas the scientific biological definition neglects the perspective of impacts and describes the naturalisation and unintended spread of unwanted organisms in areas where they have not previously occurred naturally (Jay et al., 2003; Born et al. 2005).

Organisms immigrating to new localities and their descendants have been referred to as alien, adventive, exotic, introduced and non-indigenous (Mack et al., 2000; Sharma et al., 2005).

Species that have been transported by humans from one region to another are defined as alien or exotic (Richardson et al., 2000; Villaseñor & Espinosa-Garcia, 2004).

Although there are several definitions of alien invasive species, however, the one given by GISP (2003) seems to be most relevant in totality i.e.: *'Invasive alien species are non-native organisms that cause, or have the potential to cause, harm to the environment, economies, or human health'.*

Thus establishment and spread of these species threatens landscape in terms of economy as well as environment (GISP, 2000; Sharma et al., 2005).

Established: a species with a self-sustaining population outside of its native range. Indigenous species: a species found within its native range.

Invasive species: a nonindigenous species that spreads from the point of introduction and becomes abundant.

Nonindigenous species: a species introduced to areas beyond its native range by human activity.

Noninvasive species: a nonindigenous species that remains localized within its new environment.

Transition: one step in the invasion sequence (e.g., transportation, release and establishment).

Chapter 2

Causes of Invasion

Although invasion results from multifaceted interrelated mechanisms, changes in biota resulting from habitat conversion and land use change, reduces genetic and species diversity; and due to the introduction of exotic species, leading to a homogenization of the global biota (Chapin et al., 1997). Climate and land use change also transmogrify the invasion process (Sala et al., 2000).

Land use change and habitat fragmentation may facilitate the invasion success, indirectly affecting the biodiversity. Land-use change is projected to have the largest global impact on biodiversity by the year 2100, followed by climate change, nitrogen deposition, species introductions and changing concentrations of atmospheric CO_2 (Sala et al., 2000). Land-use change is expected to be of particular importance in the tropics, climatic change is likely to be important at high latitudes, and a multitude of interacting causes will affect other biomes (Ramakrishnan, 1991; Pimm et al., 1995; Chapin et al., 2000; Lavergne et al., 2010).

Land use and biological invasions have been the major drivers of grassland biodiversity change (Zavaleta et al., 2003). Invasive species bring changes in plant community composition which may be inextricably correlated with external land-use impacts (Maskell et al., 2006). Land use change resulting from agriculture may be the major driver for invasive plants since approximately, 10^9 hectares of natural ecosystems would be converted to agriculture by 2050 (Tilman et al., 2001). Furthermore, the homogenization of Earth's biota through the establishment and spread of alien species may also be attributed to increased trade and tourism linked with globalization (Mooney & Hobbs, 2000; Rahel, 2000; Kolar & Lodge, 2002).

Table 1. Description of invasive plants investigated globally for their adverse ecological as well as economic impacts

Serial No.	Invasive plant	Impacts/Remarks	References
1.	*Lantana camara* (Figure 1A-a)	Occurs in sub-tropical and tropical ranges and destructive, both in agricultural and natural communities; considered as one of the 10 most noxious weeds in the world; rapidly invading India; alters soil physico-chemical properties	Cronk and Fuller (1995); Mack et al., (2000); Sharma et al., (2005); Totland et al. (2005); Rai (2009); Sharma and Raghubanshi (2009); Gooden et al. (2009); Sharma and Raghubanshi (2010); Osunkoya and Perrett (2011)
2.	*Centaurea maculosa* (spotted knapweed)	An economically destructive plant invader in the western United States; produce phytotoxin (–)-catechin from its roots	Callaway and Aschehoug, 2000); Bais et al. (2003); Fitter (2003); Callaway and Ridenour (2004)
3.	*Mikania micrantha* (mile a minute weed) (Figure 1A-b)	One of the top 10 worst weeds in the world	Holm et al. (1977); Zhang et al. (2004); Sharma et al. (2005); Ellison et al. (2007)
4.	*Opuntia stricta*	50-year of invasion history studied in the Kruger National Park	Foxcraft et al. (2004)
5.	*Justicia carnea*	Alien Acanthaceae species in tropical islands	Meyer and Lavergne (2004)
6.	*Odontonema strictum*	do	Meyer and Lavergne (2004)
7.	*Phlogacanthus turgidus*	do	Meyer and Lavergne (2004)
8.	*Sanchezia speciosa*	do	Meyer and Lavergne (2004)
9.	*Strobilanthes hamiltonianus*	do	Meyer and Lavergne (2004)
10.	*Hemigraphis alternate*	Form dense carpets and cover the whole ground surface	Meyer and Lavergne (2004)
11.	*Ruellia brevifolia*	The herb that colonizes the understorey of closed-canopy wet forest	Meyer and Lavergne (2004)

Serial No.	Invasive plant	Impacts/Remarks	References
12.	*Rhododendron ponticum*	Invasive shrub in North Wales	Thomson et al. (1993)
13.	*Mimosa pigra*	Reported in tropical wetland of northern Australia	Lonsdale (1993)
14.	*Baccharis pilularis ssp. consanguinea*	Reported in northern California grassland	Williams et al. (1987)
15.	*Acacia nilotica sp. indica*	Australia's worst rangeland invasive plant, introduced late last century to provide shade and feed for livestock plants	Kriticos et al. (1999)
16.	*Myrica faya*	Exotic nitrogen-fixing tree *Myrica faya* invades young volcanic sites where the growth of native plants is limited by a lack of nitrogen	Vitousek et al. (1987)
17.	*Toxicodendron radicans (L.) Kuntze*	Commonly known as poison ivy, most medically problematic invasive plants in the different parts of world	Mohan et al. (2006)
18.	*Tamarix sp.*	Dominates riparian and wetland ecosystems	Gaskin and Schall, (2002)
19.	*Solidago gigantean*	An invasive species in Europe	Jakobs et al., (2004)
20.	*Lygodium microphyllum*	Everglades ecosystem of southern Florida, USA	Volin et al. (2004)
21.	*Imperata cylindrica*	Colonizes forest lands of Asia that are cleared for slash-and-burn agriculture	Garrity et al. (1997); Chapin et al. (2000)
22.	*Trifolium* (true clover)	Recorded to be invasive in New Zealand	Gravuer et al. (2008)
23.	*Solanum mauritianum Scopoli* (Solanaceae; 'bugweed')	Major weed of natural vegetation and plantations in the eastern higher rainfall regions of South Africa	Witkowski & Garner (2008)
24.	*Impatiens glandulifera*	Invasive particularly in European regions	Kollmann and Bañuelos (2004)

Table 1. (Continued)

Serial No.	Invasive plant	Impacts/Remarks	References
25.	*Bromus tectorum, Bromus inermis*	Adapted in North America along an environmental gradient	Mack (1981); Rice & Mack (1991); Tracy et al. (2004); Kollmann and Bañuelos (2004)
26.	*Ageratina adenophora* (= *Eupatorium adenophorum* Sprengel)	Belonging to family Asteraceae, one of the worst invasive alien species in China; Also frequent in North East India	Tripathi et al. (1981); Wang and Wang (2006); Gu et al. 2007
27.	*Bryophyllum delagoense*	Major invasive species in Queensland, Australia	Witta et al. (2004)
28.	*Atriplex sagittata, A. hortensis, A. tatarica and A. rosea*	Invasive in European regions particularly central Europe	Mandak (2003)
29.	*Azolla filiculoides* (red water fern)	Native to South America, but invaded the African regions	McConnachie et al. (2003)
30.	*Cyperus rotundus* (nutgrass)	Devastating weed; secrete allelechemicals	Sharma and Gupta (2007)
31.	*Opuntia stricta* (Cactaceae)	Invasive in South African regions	Foxcroft et al., 2007
32.	*Parthenium hysterophorus L.* (Asteraceae)	Commonly known as parthenium weed, is an annual or short-lived ephemeral herb of neo-tropical origin that now has a pan-tropical distribution	Shabbir and Bajwa (2006); Timsina et al. (2011)
33.	*Fallopia japonica* (Japanese knotweed)	Causes substantial economic and environmental damage in United Kingdom (UK)	Smith et al. (2007)
34.	*Alternanthera philoxeroides (Mart.) Griseb*	Causes threat to farming systems in China	Liu-qing et al. (2007)
35.	*Lythrum salicaria*	Invasive in North America	McCaughey and Stephenson (2000)
36.	*Acroptilon repens*	Mostly confined in	Goslee et al. (2001)

Serial No.	Invasive plant	Impacts/Remarks	References
		western US and Canada; reduce forage quality, increase soil erosion and reduce wildlife populations; aggrsive vegetative growth and dominance due to water-soluble allelochemicals produced in its roots and leaves	
37.	*Campuloclinium macrocephalum* (Pompom weed)	Invasive alien plant of Grassland and Savannah biomesSouth Africa; *Liothrips tractabilis* and *Cochylis campuloclinium* are two possible biocontrol agents	Trethowan et al. (2011)
38.	*Fabiana imbricata*	Unpalatable shrub investigated in north-western Patagonia landscape; invasion could decrease grassland productivity	Ghermandi et al. (2010)
39.	*Pinus halepensis*	Investigated in mountain pampean grasslands in Argentina; horse grazing on grassland may affect the establishment of *P. halepensis*	de Villalobos et al. (2011)
40.	*Lonicera japonica* (Japanese honeysuckle)	Investigated in Cumberland Plateau and Mountain Region of USA	Lemke et al. (2011)
41.	*Pittosporum undulatum*	Woody plant invader tree or shrub) native to Australia introduced in the Azores Islands	Lourenco et al. (2011)
42.	*Prosopis juliflora*	Noxious weed particularly in dry or disturbed regions; adaptability due to tough seed coat; has got ecological and socioeconomic importance	Shiferaw et al. (2004)

Figure 1A-a. *Lantana Camara* invasion in forests of Mizoram, North-East India.

Figure 1A-b. *Mikania micrantha* invasion in invasion in forests of Mizoram, North-East India.

Figure 1A-c. *Ageratum* sp. invasion in invasion in forests of Mizoram, North-East India.

Plant invasion is now a global phenomenon. Table 1 provides the salient description of invasive plants investigated by various researchers. The invasive plants listed in Table 1 are particularly investigated in context of their ecological and economic impacts across the different continents. It is worth to mention a few prominent invasive species e.g., *Lantana camara*, *Mikania micrantha*, *Chromolaena odorata*, *Eupatorium adenophorum*, *Cytisus scoparius*, *Mimosa invisa*, *Parthenium hysterophorus* and *Prosopis juliflora* among terrestrial exotics, and *Eichhornia crassipes* and *Pistia stratiotes* among aquatics, have posed serious threat to the native flora (Sharma, 2005; Timsina et al., 2011). Henceforth, it will be better if we can provide a differential categorisation of invasion in terrestrial and aquatic ecosystems. There is a huge documentation of literature on ecosystem level impacts of plant invasion. The impacts may be further expanded on terrestrial (forest, riparian) ecosystems and aquatic (fresh water and marine) ecosystems. Ecosystem structures in terrestrial and aquatic ecosystems have many similarities; however, certain variations may have implications in context of invasion ecology. In terrestrial ecosystems, the plant-soil system involves a complex network of interactions and feedbacks and may function almost in isolation from the surrounding landscape (Ehrenfeld, 2010). Terrestrial ecosystems receive inputs from the atmosphere and from adjacent ecosystems. In contrast, aquatic systems have separate subsystems within the water column, at the water-sediment interface, and within the sediments. It is worth to mention that all these components can exchange materials with each other, as well as with other ecosystems and the atmosphere. Water movements play an important structuring role, and the dynamics of sediments are as important as the dynamics of nutrients, oxygen, and carbon in determining aquatic ecosystem function. Thus, it may be that invasives alter aquatic and terrestrial ecosystems through different pathways (Ehrenfeld, 2010).

Chapter 3

Invasion in Different Ecosystems

Invasion in Terrestrial and Riparian Ecosystems

A workshop on Impacts and Extent of Biotic Invasions in Terrestrial Ecosystems was held in Barcelona, Spain, from 19 to 22 September 2001 in order to assess the impacts and extent of biotic invasions in terrestrial ecosystems (see D'Antonio & Kark, 2002). In the aforesaid meet, many ways were identified in which future research can help to address unresolved questions, particularly by the expansion of the regions and type of species being studied (Long, 1981; Lever, 1987; Williamson & Fitter, 1996; Lonsdale, 1999; Blondel & Aronson, 1999; Parker et al., 1999; D'Antonio. & Kark, 2002). Invasive plants also affect the forest succession. Anthropogenic perturbations linked with forestry and agriculture such as logging and subsistence agriculture may promote the establishment of nonnative, invasive plant species, potentially affecting forest structure and diversity even long after the perturbation has ceased (Brown & Gurevitch, 2003). Brown and Gurevitch (2003) investigated the effects of limited logging on the presence, persistence, and impact of invasive species on forest composition in Ranomafana National Park in southeastern Madagascar, a biodiversity hotspot. Their (Brown & Gurevitch, 2003) results confirmed that invasive plants are not transient members of postlogging tropical forests in Madagascar but maintain long-term viable populations after their initial colonization and can dramatically shift the trajectory of forest succession.

Lantana camara is an important weed of agro and forest ecosystems, where it forms dense thickets that livestock cannot penetrate. The leaves are toxic when ingested by most domestic livestock or native mammals, although toxicity varies greatly between strains (Goulson & Derwent, 2004). Tropical forest biodiversity is declining and results indicated that future carbon storage in tropical forests will be influenced strongly by future species composition (Bunker et al., 2005).

Riparian zones are among the most diverse habitats on earth; however, they are also heavily invaded by alien plants like terrestrial ecosystems (Naiman & Décamps, 1997; Stohlgren et al., 1999; Greenwood et al., 2004; Yong Sunga et al., 2011). Riparian ecosystems are particularly prone to plant invasions due to their position in lower lying areas in the landscape and the relatively high frequency and often severity of disturbances such as floods (Holmes et al., 2005; Richardson et al., 2007; Beater et al., 2008). In such ecosystems, integrating clearing with restoration of the tall indigenous riparian canopy tree species in heavily invaded sites would help to shade out many alien recruits (Beater et al., 2008). In a California riparian system, the most diverse natural assemblages are the most invaded by exotic plants (Levine, 2000). Woody plant invaders, such as *Tamarix* spp. and *Melaleuca quinquenervia* in North America, and *Acacia mearnsii* in South Africa (Richardson et al., 1997; Di Tomaso, 1998; Serbesoff-King, 2003; Greenwood et al., 2004) are now among the worst invaders of riparian zones (Mack et al., 2000; Hood & Naiman, 2000; Greenwood et al., 2004). Greenwood et al. (2004) changes in abundance, diversity, and composition of terrestrial arthropods following *Salix rubens* (willow) invasion of the riparian zone may indirectly alter in-stream food webs and have important effects on higher-order consumers in the riparian zone. *Impatiens* reduces native species diversity in open and frequently disturbed riparian vegetation, particularly widespread ruderal species (Hulme & Bremner, 2006). Yong Sunga et al. (2011) examined twelve riparian forests along urban–rural gradients in Austin, TX (USA), on the relationship among watershed urbanization and the invasion of alien woody species concluded that watershed urbanization facilitates the invasion of alien species in riparian forests by causing hydrologic drought, particularly in hot and semi-arid regions.

Invasion in Aquatic Ecosystems

In previous section, I have discussed the invasion of terrestrial and riparian ecosystems; however, discussion on the extent of invasion will be rather incomplete, if the aquatic ecosystems will remain untouched. Invasion by mats of free-floating plants is among the most important threats to the functioning and biodiversity of freshwater ecosystems ranging from temperate ponds and ditches to tropical lakes (Scheffer et al., 2003). In the case of *Phragmites australis*, anthropogenic linear wetlands such as roadside and agricultural ditches are believed to play a key role in invasion patterns (Maheu-Giroux & de Blois, 2005). *Hydrilla verticillata* is an invasive submerged weed, was investigated for isozyme analysis and inferred that reproduction is solely by vegetative means (Hofstra et al., 2000). Dark, anoxic conditions under thick floating-plant cover leave little opportunity for animal or plant life, and they can have large negative impacts on fisheries and navigation in tropical lakes (Scheffer et al., 2003).

Biological invasions are not confined to fresh water only; rather, marine ecosystems are also being affected. During the past decade, a rapid increase in studies of coastal invasions has provided important insight into the invasion process in marine systems and how these invasions might differ qualitatively from invasions occurring in terrestrial and fresh water systems (Grosholz, 2002).

Mollo et al. (2008) adopted a chemo-ecological approach in order to define biotic conditions that enhance biological invasions in terms of enemy escape and resource opportunities. Mollo et al. (2008) further focused on the secondary metabolite composition of three exotic sea slugs found in Greece that have most probably entered the Mediterranean basin by Lessepsian migration and concluded that initial invasion of this exotic pest would seem to have paved the way for the subsequent invasion of a trophic specialist that takes advantage of niche opportunities. Invasive species also tend to alter the interaction webs as demonstrated in a case study of marine community (Hollebone & Hay, 2007).

The control of alien marine species is in its infancy (Bax et al., 2001). In coastal waters of the U.S., >500 invaders have become established, and new introductions continue at an increasing rate (Grosholz, 2005). Through the lab and field experiments on introduced crab, Grosholz (2005) suggested that positive interactions among the hundreds of introduced species that are accumulating in coastal systems could result in the rapid transformation of previously benign introductions into aggressively expanding invasions.

Chapter 4

Global Problems of Invasion: Continent-Wise Impact

Invasive plant species threaten the integrity of natural systems throughout the world by displacing native plant communities (Kennedy et al., 2002) and establishing monocultures in new habitats (Callaway, 2002; Bais et al., 2003). The invasive species composition of a country reflects historical as well as recent processes (Dehnen-Schmutz, 2004).Non-native species now dominate most landscapes in most parts of the world (Didham et al., 2005). Prompt role of media as well as popular articles has also provoked a awareness and discussion about the global impacts of invasive species (Simberloff, 2004). This necessitated the urgent global scientific researches in order to have concrete policies to curb the adverse impacts originating from plant invasion (Rosenzweig, 2001; Slobodkin 2001; Gurevitch & Padilla, 2004; Didham et al., 2005).

Continent wise, Europe is reported to be the source of many of the world's worst invasive species, including Austrian pine (*Pinus nigra*), Norway maple (*Acer platanoides*), Spanish slug (*Arion lusitanicus*), German wasp (*Vespula germanica*), Scotch broom (*Cytisus scoparius*), and English starling (*Sturnus vulgaris*) (Hulme et al., 2009). *Atriplex sagittata, A. hortensis, A. tatarica* and *A. rosea* are annual heterocarpic early succesional species not native in Central Europe (Mandak, 2003).

In Australia (Leigh & Briggs, 1992; Groves & Willis, 1999), like other countries, plant invasion has been associated with the extinction of several valuable endemic plant species (Briggs & Leigh, 1996) like *Lantana camara* (Gooden et al., 2009). Threatened endemic species may be particularly vulnerable to plant invasion due to their small population sizes, poor competitiveness or restricted geographic range (Walck et al., 1999). McDougall et al. (2005) recorded 128 invasive plant species in treeless vegetation of the Australian Alps

Australia and New Zealand have evolved plant biodiversity with a high degree of endemism, and which is particularly susceptible to environmental weeds (Williams & West, 2000). In Australia alone, *L. camara* is currently estimated to cover c. 40,000 km^2 (Goulson & Derwent, 2004). There are now as many alien established plant species in New Zealand as there are native species (Mooney & Cleland, 2001). It has been reported that many countries have 20% or more alien species in their floras (Vitousek et al., 1996). There are few geographic generalities to these trends; the strongest is that islands, in particular, have been the recipients of the largest proportional numbers of invaders (Mooney & Cleland, 2001). Biotic homogenization within continents is equally as striking as mixing among oceans (Mooney & Cleland, 2001). Rahel (2000) noted that in the United States pairs of states on average now share 15 more species than they did before European settlement. The states of Arizona and Montana, which previously had no fish species in common, now share 33 species in their faunas. Mack (1985) estimateed that over the last 500 years, invasive species have come to dominate 3% of the Earth's ice-free surface. Vast land or waterscapes, in certain regions, are completely dominated by alien species, such as the star thistle *Centaurea solstitialis* in the rangelands of California, cheatgrass (*Bromus tectorum*) in the intermountain regions of the western United States, and water hyacinth (*Eichornia crassipes*) in many tropical lakes and rivers (Mooney & Cleland, 2001). Cohen and Carlton (1998) noted that the rate of invasion into San Francisco Bay has increased from approximately one new invader per year in the period of 1851–1960, to more than three new invaders per year in the period of 1961–1995.

In relation to invasive species it is worth to mention that among various continents, Africa is still recognised as 'Dark Continent' (Pimm, 2007). The hotspots of Africa are vulnerable to invasive species and unlike other places South Africa has a particularly active program for removing them to restore ecosystem services (Pimm, 2007). *Opuntia stricta* (Cactaceae), an alien weed, has invaded an area of more than 35,000 ha in the Skukuza region of the Kruger National Park, South Africa and is difficult to manage as its invasion is independent of environmental factors, revealed by Canonical Community Analysis (Foxcroft et al., 2007). Ashanti region of Ghana, S. Africa is heavily invaded by exotics like *Chromolaena odorata* (L.) King & Robinson, *Centrosema pubescens* Benth. and *Rottboellia cochinchinensis* (Lour.) Clayton etc. which affect agriculture, forestry and economy, hydropower generation and drinking water supply systems. Invasive species adversely affect the biodiversity in conjunction with intensive agriculture and rapid urbanization as demonstrated in Cape Floristic Region, South Africa (Rougeta, 2003).

A list of 618 species of alien flowering plants recorded for Mexico (Villaseñor & Espinosa-Garcia, 2004). *Lythrum salicaria* is an invasive species in eastern North America (McCaughey & Stephenson, 2000). Invasion by exotic species has affected desert ecosystems in southwestern US and northwestern Mexico and is considered one of the main threats to biodiversity in the Sonoran Desert (Van Devender et al., 1997). *Azolla filiculoides* (red waterfern) is a floating fern native to South America which has invaded aquatic ecosystems in South Africa (McConnachie et al., 2003). Over 120,000 non-native species of plants, animals and microbes have invaded the United States, United Kingdom, Australia, South Africa, India, and Brazil, and many have caused major economic losses in agriculture and forestry as well as negatively impacting ecological integrity (Pimentel et al., 2001).

Likewise, problem of invasive plant species is growing at an alarming rate in Asian countries like India and China. *Lantana camara, Mikania micrantha, Chromolaena odorata, Eupatorium adenophorum, Cytisus scoparius, Parthenium hysterophorus* and *Prosopis juliflora* (terrestrial exotics), and *Eichhornia crassipes* and *Pistia stratiotes* (aquatic exotics) are some of the prominent invasive plants reported from India (Sharma et al., 2005). *Prosopis juliflora, Capparis seplaria, Grewia flavescens* and *Lantana Camara* are the dominant invasive species in many parts of Rajasthan resulting from landscape modernization (Robbins 2001). Invasive species are growing at an alarming rate in some ecologically relevant regions of India like e.g., Sundarbans (Biswas et al., 2007). Crofton weed (*Ageratina adenophora*) is a highly invasive weed that has spread into several provinces of southern China (Gu et al., 2007). Wu et al. (2004) compiled a list of casual and naturalized species and found that about 60% of exotic species were probably introduced unintentionally into Taiwan; many species imported intentionally have ornamental, medicinal, or forage values. *Alternanthera philoxeroides* (Mart.) Griseb, an invasive weed, causes threat to farming systems in China (Liu-qing et al., 2007**).** *Cyperus brevifolius* and *Cyperus kyllingia* have invaded most parts of Indonesia which may be attributed to their adaptability to varying soil types (Rodiyati et al., 2005).

Island biotas, as demonstrated in Australia and New Zealand, are particularly vulnerable to disruption by the spread of alien plants due to the restricted range and small population size of its native species and habitats

(Meyer & Lavergne, 2004). Islands countries particularly those existing in tropics are severe victims due to multiple cases of invasion by alien plant species that have severely disturbed their natives. These disturbances have occurred at the ecosystem level (through the alteration of hydrological and fire regimes, changes in the nutrient cycles, and increased soil erosion), the community level (through the decline or loss of food sources, host-plants, and nesting sites for native animals), and the species level (through the local extirpation of native plants populations, sometimes leading to extinction) (Meyer & Lavergne, 2004). La Réunion Island belonging to Indian Ocean is identified to be threatened one from biological invasions from *Fuchsia magellanica, Furcraea foetida, Hiptage benghalensis, Clidemia hirta, Strobilanthes hamiltonianus, Ulex europaeus* and *Acacia heterophylla* forest. Australian acacias, which are invasive in western Indian Ocean islands, were introduced in beginning of the nineteenth century (Kull et al., 2007). In Galapogas Island, several invasive plant species e.g., *Rubus niveus, Cestrum auriculatum, Pennisetum purpureum, Cedrela odorata* were reported to be threat to native biodiversity of this pristine ecosystem (Mauchamp, 1997).

Apart from several globalized countries/continents, invasion problem may now touch the pristine ecosystems of Antarctica. Invasions are widely recognized to constitute a serious risk to the Antarctic region (Pugh, 1994; Dingwall, 1995; Smith, 1996; Chown et al., 2001; Greenslade, 2002; Frenot et al., 2005), With rapid climate change *occurring* in some parts of Antarctica, elevated numbers of introductions and enhanced success of colonization by aliens are likely, with consequent increases in impacts on ecosystems (Frenot et al., 2005).

Chapter 5

Impacts of Invasion

Plant invasion is now intimately linked with environment and economy. There is an increasing realization of the ecological costs of invasion process and there is an urgent necessity to establish the linkage between invasion and native biodiversity.

Invasive plants can alter plant community structure and ecosystem function (Vitousek et al., 1987), result in large economic costs from lost ecosystem services (Pimentel et al., 2005), and detract from an intrinsic or aesthetic value associated with native biodiversity and native plant dominance. Identifying suites of plant traits and corresponding ecological strategies used by successful invaders would improve our understanding of how particular species and landscape features interact to produce the explosive spread of invasive species. Accurate assessment of the invasive potential of an introduced species before introduction would provide a valuable tool to reduce invasions.

Ecological/Environmental Impacts

Ecosystems deliver a wide range of services to human society (e.g., Daily, 1997; Constanza et al., 1997; van Wilgena, 2007). A 4-year global assessment of the world's ecosystem services (Millennium Ecosystem Assessment, 2005) found that 60% of the services assessed were declining in condition due to a suite of anthropogenic drivers (such as habitat loss and alteration, water abstraction, overexploitation, and invasive alien species (van Wilgena, 2007).

The invasion of ecosystems by alien species has been identified as a large and growing threat to the delivery of ecosystem services (Drake et al., 1989; van Wilgena, 2007). The most pertinent consequences of this reshuffling is a sharp increase in invasive alien species that establish new ranges in which they proliferate, spread, and persist at the cost of native species (Mack et al., 2005).

Furthermore, invasive plants cause noticeable alteration to the composition, structure, and functioning of ecosystems (Cronk & Fuller. 1995; Adair & Groves, 1998; Yeates & Williams, 2001). Plant invasion is coupled with large ecosystem-level impacts such as changing soil nitrogen regimes (J. Egunjobi, 1971; Vitousek & Walker, 1989; Yeates & Williams, 2001; Sharma & Raghubanshi, 2009), increased fire frequency and changed vegetation microclimate (D'Antonio & Vitousek, 1992; Wardle et al., 1995; Yeates & Williams, 2001), and allelopathic effects (Gentle & Duggin, 1997; Yeates & Williams, 2001).

Likewise, plant invasion in an integrated complex way perturb the ecosystem based processes (Adair, 1995; Hobbs & Humphries, 1995; Vitousek et al., 1996; Adair & Groves 1998; DiTomaso, 2000; Chapin et al., 2000; Levine et al., 2003; Dukes & Mooney, 2004; D'Antonio & Hobbie, 2005; Theoharides & Dukes, 2007). The negative impacts, that invasive plant have on ecosystem structure and functioning, means that their presence is often incompatible with the ideals of sustainable management or conservation (Higgins & Richardson 1996). However, quantifying the environmental damage and loss of biodiversity due to alien species invasions worldwide is complicated by the fact that some 1.5 million species of the estimated 10 million species on earth are identified and described (Raven & Johnson, 1992; Pimentel et al., 2001). Invasive plants displace native plants (McKnight, 1993; Groves & Willis, 1999) and change the composition of native plant communities (Chapin et al., 2000). Invasive plants alter the ecosystem processes such as nutrient cycling (Vitousek & Walker, 1989; Vitousek, 1990) and disturbance regimes (Vitousek, 1990; Mack & D'Antonio, 1998). Changes in the abundance of species, especially those that influence water and nutrient dynamics, trophic interactions, or disturbance regime, affect the structure and functioning of ecosystems (Chapin et al., 1997). Interestingly, grasses constitute a major group of invasive plants that can dramatically alter native plant community structure and ecosystem processes such as fire frequency, nutrient cycling, and water circulation (Lavergne & Molofsky, 2007). Further, integral components of ecosystems like pollinators and insects may be adversely affected by the dominant invasives (deGroot et al., 2007).

Invasive weeds may have a wide range of consequences in pasture ecosystems as demonstrated in case of *S. jucobueu* (Wardle et al., 1995). At Mont Saint-Michel bay of France invasion of *Elymus athericus* belonging to Poaceae has drastically changed its vegetation and pertaining to their invasion spider population acted as a bioindicator (Pe´tillon et al., 2005).

Global change biology is largely driven by change in the vegetation pattern brought through invasive plant species (Mack et al. 2000, Sharma et al. 2005). The invasion of exotic plant species is an important component of global environmental change (Theoharides & Dukes, 2007).

Over 40% of the species on the list of threatened and endangered species is due to invasive species (Wilcove et al., 1998; Sharma et al., 2005). Rejmanek and Randall (1993) estimated that 20% or more of the plant species is non-indigenous in many continental areas and 50% or more on many islands. The impacts of invasive species can be more pervasive than simple reduction of species numbers (Sanders et al., 2003). Invasive species not only reduce biodiversity but rapidly disassemble communities and, as a result, alter community organization among the species that persist (Sanders et al., 2003).

Plant invasions by exotic species can also alter the composition of natural and agricultural communities substantially (Vitousek et al., 1987). Biological invasions by exotic species can also alter fluxes of energy, water, and nutrients could be similarly valuable, therefore, whole ecosystems can be regulated by the populations and properties of individual species (Vitousek et al., 1987). Exotics can affect disturbance regimes, such as the familiar examples of increased erosion attributed to introduced herbivores such as goats, sheep, and rabbits (Coblentz, 1978; Chapuis et al., 1994).

Invasive plants or pathogens have some of the most dramatic effects on forest ecosystems. Pathogens that influence dominant canopy species include aboveground pathogens, such as the ascomycete, *Cryphonectria parasitica*, attacking American chestnut, *Castanea dentata*, in the eastern US, and invasive root pathogens, such as *Phytophthora lateralis* attacking Port Orford cedar (*Chamaecyparis lawsoniana*) in the northwestern US (Jules et al., 2002), or P*hytophthora cinnamomi* causing dieback in eucalypt forests in western Australia. In the case of *Phytophthora cinnamomi t*he dieback of susceptible eucalyptus trees leads to shifts in plant and animal communities both above and belowground (Newell, 1998; Postle et al., 1986; Weste et al., 2002).

Vitousek et al. (1987) studied ecosystem-level consequences of invasion by *Myrica faya* (a small tree with a nitrogen-fixing symbiosis) into ecosystems developing in young volcanic substrates in Hawaii Volcanoes National Park. Such sites contained no native plants with nitrogen-fixing symbioses. The exotic nitrogen-fixing tree *Myrica faya* invades young volcanic sites where the growth of native plants is limited by a lack of nitrogen. *Myrica* quadruples the amount of nitrogen entering certain sites and increases the overall biological availability of nitrogen, thereby altering the nature of ecosystem development after volcanic eruptions (Vitousek et al., 1987).

Mueller-Dombois (2000) reviewed establishment and succession of tropical rain forest vegetation dominated by single canopy *Metrosideros polymorpha* (Myrtaceae) in the post-mining landscapes of Hawaii. This species appears as the first tree on raw volcanic substrates and maintains its dominant canopy position by a growth cycle involving canopy dieback and then it creates secondary successions favorable for its own regeneration as a shade-intolerant pioneer tree (Mueller-Dombois, 2000).

Pfisterer et al. (2004) indicated that the positive diversity-productivity relationships during the weeding phase were mainly regulated by species richness. Zavaleta and Hulvey (2004) investigated observed variation in grassland diversity to design an experimental test of how realistic species losses affect invasion resistance and predicted that realistic biodiversity losses, even of rare species, can thus affect ecosystem processes far more than indicated by randomized-loss experiments.

There may be differential impacts of invasion on wildlife. Top predators with their large body size, low abundance, and large home range requirements are particularly vulnerable to habitat fragmentation and hence invasion (Raffaelli, 2004). In contrast, because of the higher species richness at lower trophic levels, there is more *"insurance"* against the effects of species losses, and these species may also have a greater capacity for adaptive change due to their shorter life-spans and faster turnover rates. With respect to species traits, extinction is unlikely to occur randomly (Raffaelli, 2004)

There exists scarce work pertaining to the impacts of invasive weeds on smaller organisms, particularly those which are major drivers of below-ground ecosystem processes and focus in this regard is warranted as they respond rapidly to environmental perturbations, making them good indicators of changing conditions (Ingham et al., 1985; Edwards 2000; Yeates & Boag, 2001; Yeates & Williams, 2001).

Economic Impacts

Despite disrupting ecosystems level processes, invasive species compete with native species and cause huge global economic losses. Some invasive species have caused major economic losses in agriculture, forestry, and several other segments, in addition to harming the environment (Pimentel et al., 2000). Native biodiversity and its links to ecosystem properties have cultural, intellectual, aesthetic and spiritual values that are important to society (Chapin et al., 2000). In addition, changes in biodiversity that alter ecosystem functioning have economic impacts through the provision of ecosystem goods and services to society (Chapin et al., 2000). These impacts can be wide-ranging and costly. For example, the introduction of deep-rooted species in arid regions reduces supplies and increases costs of water for human use. Marginal water losses to the invasive star thistle, *Centaurea solstitialis*, in the Sacramento River valley, California, have been valued at US$16–56 million per year (Chapin et al., 2000). In South Africa's Cape region, the presence of rapidly transpiring exotic pines raises the unit cost of water procurement by nearly 30% (Van Wilgen et al., 1996, Chapin et al., 2000)). Thus we can say that invasive species have got societal impacts also.

Increased evapotranspiration due to the invasion of *Tamarix* in the United States costs an estimated $65–180 million per year in reduced municipal and agricultural water supplies. In addition to raising water costs, the presence of sediment-trapping *Tamarix* stands has narrowed river channels and obstructed over-bank flows throughout the western United States, increasing flood damages by as much as $50 million annually (Zavaleta, 2000; Chapin et al., 2000).

Those species changes that have greatest ecological impact frequently incur high societal costs (Chapin et al., 2000). Changes in traits maintaining regional climate (Shukla et al. 1990, Chapin et al. 2000) constitute an ecosystem service whose value in tropical forests has been estimated at $220 ha^{-1} yr^{-1} (Costanza et al., 1997; Chapin et al., 2000).

Pimentel et al. (2000) mentioned the cost benefit analysis of invasive species (both vertebrates and invertebrates) in US. The approximately 50,000 non-indigenous species in the United States cause major environmental damage and losses totalling approximately $137 billion per year (Pimentel et al., 2000). Large economic impacts are also associated with many invasive species, which can provoke agricultural losses, disrupt ecosystem services, and lead to disease proliferation (Andow et al., 1990, Arim et al., 2006). Arim et al. (2006) investigated that spread of invading birds, amphibians, invertebrates, fish, plants, and a virus exhibited similar negative feedback structures that stabilized their rate of spread and hence regulation in invasion advancement is a widespread phenomenon in nature and regulation structure is strikingly consistent among invasions. Through their experiments, Arim et al.

(2006) opposed current beliefs that species invasions are idiosyncratic phenomena, and general patterns do exist. In the USA alone, the estimated economic impact of invasive species is $34 billion per year (Pimentel et al., 2005, Theoharides. & Dukes, 2007). Over the past few centuries, thousands of species of freshwater, estuarine, and marine organisms have dispersed outward from their native regions through human-mediated transport and have established sustaining populations in distant parts of the globe (Elton, 1958; Walford & Wicklund, 1973; Cohen & Carlton, 1998) at the cost of many economic plants, animals and microbes. Many of these organisms have profoundly affected the abundance and diversity of native biota in the regions they have invaded (Zibrowius, 1991; Leppa¨ koski, 1993; Cohen & Carlton, 1998), and in some cases they have had substantial economic impacts (Nalepa & Schloesser, 1993; Cohen & Carlton, 1998).

Some invasive plants indirectly lead to economic losses e.g., *Imperata cylindrica*, an aggressive indigenous grass, colonizes forest lands of Asia that are cleared for slash-and-burn agriculture, forming monoculture grassland with no vascular plant diversity and many fewer mammalian species than the native forest (Chapin et al., 2000). The total area of *Imperata* in Asia is currently about 35 million ha (4% of land area) (Garrity et al., 1997; Chapin et al., 2000).

Chapter 6

Diversity-Stability-Invasibility Dispute

Global researches established that the intrusion of alien species has negative effects on the existing plant community, leading to reduced diversity and the loss of native biodiversity (Levine et al., 2003; Brooker 2006). Thus, it may be assumed that diversity and stability of an ecosystem may have inextricable linkage with invisibility. The relationship between diversity and stability has been a matter of great dispute among ecologists since the inception of the discipline (Elton, 1958; Ives & Carpenter, 2007). As per diversity-stability hypothesis/relationship in ecology stating that complex natural communities are more stable than simple ones has been contested widely in literature (Goodman, 1975).

Diversity is also functionally important, as it increases the probability of including species that have strong ecosystem effects. Differences in environmental sensitivity among functionally similar species give stability to ecosystem processes, whereas differences in sensitivity among functionally different species make ecosystems more vulnerable to change (Chapin et al., 1997). Current global environmental changes that affect species composition and diversity are therefore profoundly altering the functioning of the biosphere (Chapin et al., 1997).

Furthermore, species-rich systems were more likely to show a greater range of diversity-stability relationships; the prevalence of systems with alternative stable states and nonpoint attractors increased with diversity (Ives & Carpenter, 2007).

It has been demonstrated that in experimental communities of sessile marine invertebrates, increased species richness significantly decreased invasion success, apparently because species-rich communities more completely and efficiently used available space, the limiting resource in marine systems (Stachowicz, 1999). Declining biodiversity thus facilitates invasion in this system, potentially accelerating the loss of biodiversity and the homogenization of the world's biota (Lodge, 1993; Stachowicz, 1999). However, in case of wetlands in Ontario, it has been demonstrated that exotic species richness was positively related with native ones and therefore, it was recommended to prevent the spread of community dominants regardless of geographical origins (Houlahan & Findlay, 2002).

In coastal grasslands of south-eastern Australia, disturbances like grazing and burning triggered invasion by the indigenous shrub, *Acacia sophorae* and there was strong negative correlation between *Acacia* cover and native species richness (Costello et al., 2000).

Whether Species Richness Prevent Invasion?

Several recent invasion experiments considering species diversity as the treatment variable supported the prediction that high diversity communities are more resistant to invasion (Tilman 1993, 1997; Levine, 2000; Kennedy et al., 2002. Huston, 2004). However, other experiments demonstrated no effect of species richness on invisibility (Lavorel et al., 1999). Re-evaluation of the positive experimental results suggested that factors apart from variation in species diversity, such as total biomass or plant density, may actually be causing the observed responses (Wardle, 2001; Weltzin et al., 2003; Huston, 2004). Henceforth, manipulation of productivity and disturbance paves the way for resource managers to influence the interactions among species in order to eliminate some types of invasive species (Huston, 2004).

Biodiversity can affect the ability of exotic species to invade communities through either the influence of traits of resident species or some cumulative effect of species richness (Chapin et al., 2000). Early theoretical models and observations of invasions on islands indicated that species-poor communities would be more vulnerable to invasions because they offered more empty niches (Elton, 1958; Chapin et al., 2000). However, studies of intact ecosystems find both negative (Tilman, 1997) and positive (Levine & Antonio, 1999) correlations between species richness and invasions.

Biodiversity can also impact the resistance of communities to invasion (Purvis & Hector, 2000). Although exceptions exist, in experiments which manipulate diversity directly, communities with more species are often more resistant to invasion probably for the same reason that they are more productive (Purvis & Hector, 2000). While studying on invasion ecology of porcelain crab (*Petrolisthes armatus*) in oyster reefs of Georgia, USA, Hay and Hollebone (2007) demonstrated that biotic resistance from native species richness slows initial invasion, but early colonists paves the way for later ones and produce tremendous propagule pressure which offshoot the effects of biotic resistance.

Diversity of one group of organisms can also promote diversity of associated groups, for example between mycorrhizas and plants or plants and insects (Van der Heijden et al., 1998; Stachowicz et al., 1999; Knops et al., 1999; Schwartz et al., 2000; Purvis & Hector, 2000). However, recent studies show that the diversity of exotic species is often positively correlated with native plant species diversity (e.g., Lonsdale, 1999; Stohlgren et al., 1999), which contradicts this perception and suggests that high native species diversity by no means functions as a barrier to exotic invasion. In addition, anthropogenic disturbance that create openings in tropical forest habitats makes them susceptible to invasion by exotic species. For example, in tropical forest-reserves, trails are kept open by removing bordering vegetation to allow eco-tourists and others to experience tropical forest habitats in a convenient way, resulting in the formation of gaps of contrasting sizes along such trails. Exotic populations establishing in these gaps may serve as starting points from where seeds may disperse into pristine forest interiors or naturally disturbed patches within the forest (Murcia, 1995; Cadenasso & Pickett, 2001, but see Makana & Thomas, 2004).

Chapter 7

Ecological Attributes
and Strategies of Invasion

In previous sections of this review, we observed that anthropogenic perturbations have caused an unprecedented redistribution of the earth's flora and fauna (Mack et al., 2000; 2003). As a result of the rapid land use changes, the pace of invasion has particularly accelerated during the past century (Schei, 1996; Sharma et al. 2005). However, invasive plants are actually like native plants, not bad by birth, is just a matter of being in a favourable environment or possessing certain ecological attributes which triggers them as invasive ones. It is rather impossible to predict the ecological behaviour of a species in a new environment (Anon, 1998; Williamson, 1999). Species whose native status and origin are not clear is called cryptogenic species (Carlton, 1996; Sharma et al. 2005).

Quest for the ecological mechanism lying behind the success of invasive species over native species has drawn the attention of researches worldwide (Keane & Crawley, 2002; Williams et al., 1995; Totland et al., 2005). In this regard, ecologists are trying to investigate the prime factors that determine plant abundance (Van der Putten, 2002).

Transport, colonization, establishment and landscape spread may be different steps in success of invasive plants and each and every step is checked through several ecological attributes (Figure 2). Transport of invasive plants in the form of seed/seedlings is checked through propagule pressure necessary for colonization which inturn is checked by different abiotic factors (like temperature, sunlight, moisture etc.). Lockwood et al. (2005) reviewed that propagule pressure is a key element to understanding why some introduced populations fail to establish whereas others succeed. Even if the exotic plants are able to colonize, native plants offer biotic resistance for their establishment. After escaping biotic resistance, invasive plants spread rapidly across the landscape. Landscape spread is also constrained through habitat connectivity and dispersal ability. Moreover, all these steps are inextricably linked with global environmental change, fire/disturbance regime and extinction of native biodiversity through land use change or habitat fragmentation (Figure 2).

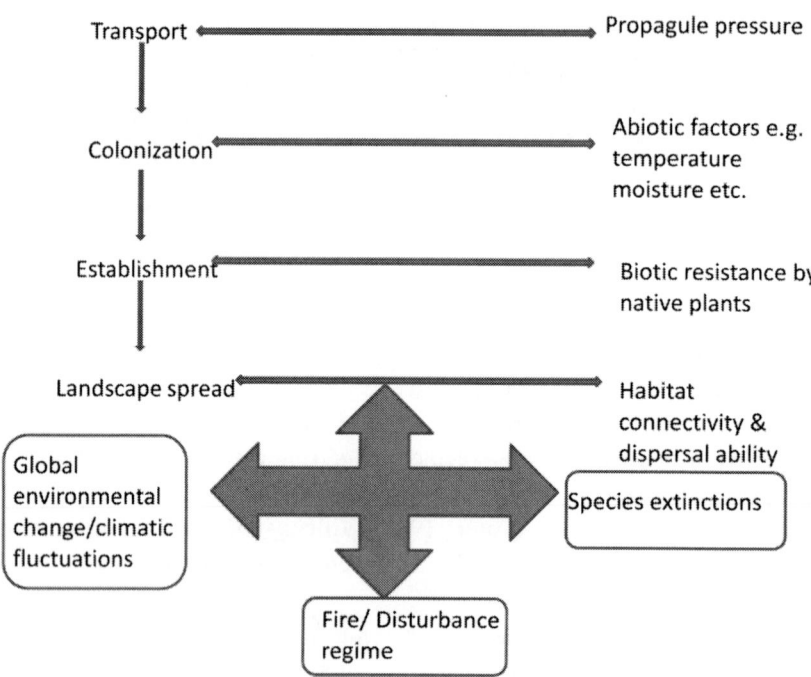

Figure 2. Filters or checkpoints and factors affecting different steps of invasion [Modified after Chapin et al. (2002); Theoharides and Dukes (2007)].

Davis et al. (2000) opined that the elusive nature of the invasion process arises from the fact that it depends upon conditions of resource enrichment or release that have a variety of causes but which occur only intermittently and, to result in invasion, must coincide with availability of invading propagules. The actual invasion of an environment by new species is influenced by three factors i.e., the number of propagules entering the new environment (propagule pressure), the characteristics of the new species, and the susceptibility of the environment to invasion by new species (invasibility) including the region's climate, the environment's disturbance regime, and the competitive abilities of the resident species (Lonsdale, 1999; Davis et al., 2000).

To understand the factors that determine plant invasiveness and competitive ability (Field et al., 2006) is a major challenge in plant biology. The threat posed by non-native invasive plant species has spurred efforts to identify individual species that show a high probability for naturalization and/or invasiveness and to rapidly eradicate those species while their distributions are limited or prevent their introduction. However, there is a widespread perception that predictions about which species will invade are impossible (Enserink, 1999). Colautti and MacIsaac (2004) synthesized an invasional framework based on current models that break the invasion process into a series of consecutive, obligatory stages. Under their framework, invasions can be more objectively understood as biogeographical, rather than taxonomic, phenomena (Colautti & MacIsaac, 2004).

In order to define the plant-environment interaction, it is useful to consider the environment as a series of filters which prevent unsuited plants from establishing, maturing, reproducing and dispersing (Keddy, 1992). The selection of the appropriate autecological attributes and environmental filters requires an understanding of how the system functions. (Higgins & Richardson, 1996). Both the biotic and abiotic properties of the target habitat are likely to be as important as the aut-ecological attributes of the invading species in influencing invasive success (Higgins & Richardson, 1996).

Inextricable linkage and interactions between climate change, species invasions, and habitat fragmentation could cause further diversity losses, because many species may be unable to migrate through fragmented habitats to reach regions with suitable climates and soils (Sala et al., 2000; Tilman et al., 2001; Crossman et al., 2011).

Invasion Window Concept

The *"invasion window"* concept (Johnstone, 1986) emphasizes the importance of the temporal availability of resources in influencing invasive success. Equally important, however, is the spatial pattern of resource availability. Resources in terms of resource availability and disturbance play an important role in the process of plant invasion. Resource availability includes factors such as nutrient, moisture and space availability which can be regarded as manifestations of spatial and temporal environmental heterogeneity. The importance of environmental heterogeneity in influencing invasion patterns is well established. For example, Lonsdale (1993) observed a strong correlation between the areal spread of the invasive shrub *Mimosa pigra* in a tropical wetland of northern Australia and the previous year's rainfall. Similarly, Williams et al. (1987) noted that establishment of *Baccharis pilularis* ssp. *consanguinea* in northern California grassland was correlated with annual and spring rainfall. In another example, nutrient availability strongly influenced the invisibility of Californian serpentine grassland by alien annual grasses (Hobbs et al., 1988; Huenneke et al., 1990). Both biotic and abiotic attributes of the environment were experimentally shown to govern the invasibility of a Californian coastal plant community (D'Antonio, 1993). The same was illustrated more indirectly in a multiple regression analysis which related a number of habitat and land-use attributes to the distribution and spread of the invasive shrub *Rhododendron ponticum* in North Wales (Thomson et al., 1993).

Phenotypic Plasticity

In plants, the well developed plasticity of many traits is usually interpreted as an adaptive response to environmental heterogeneity as a consequence of immobility and modular growth (Dorken & Barrett, 2004). Although studies of phenotypic plasticity have a long history in plant ecology (Bradshaw, 1965; Schlichting, 1986; Scheiner, 1993; Schlichting & Pigliucci, 1998; Pigliucci, 2001; Dorken & Barrett, 2004), the extent to which patterns of plasticity differ among traits, life histories and habitats, and the adaptive basis of this variation are largely unresolved questions.

In the C_4 African grass, *Pennisetum setaceum*, on the other hand, phenotypic plasticity was more important than local adaptation to dominance across diverse habitats on Hawaii (Williams et al., 1995; Kollmann & Bañuelos, 2004); similar results were reported for the invasive alien *Agrostis capillaries* in New Zealand (Rapson & Wilson, 1992; Kollmann & Bañuelos, 2004).

Dorken & Barrett (2004) investigated the plasticity of vegetative and reproductive traits in *Sagittaria latifolia*, a clonal aquatic plant whose populations are both monoecious and dioecious. They (Dorken & Barrett, 2004) evaluated the prediction that populations of the two sexual systems would have different patterns of phenotypic plasticity because of the contrasting habitats in which they occur and found significant plasticity for female sex allocation in monoecious populations, with more female flowers at higher nutrient levels. Järemo & Bengtsson (2011) shows that the organism life-history may control the effect of age of introduced individuals on the probability of establishment of a new population and that competition has a larger effect on semelparous organisms than iteroparous. Therefore, life history traits and age structure may also play an important role in invasion process (Järemo & Bengtsson, 2011).

Kollmann & Bañuelos (2004) described variations in growth and phenology in 26 populations of *Impatiens glandulifera* from nine European regions in a common garden in Denmark. They (Kollmann & Bañuelos, 2004) described the potential consequences of such latitudinal trends for population dynamics and dispersal of alien plants. Small colonizing populations generally have increased rates of evolution, and that might be particularly true in invasive alien species (Eckert et al., 1996; Lee, 2002; Kollmann & Bañuelos, 2004). One other example is the European grass *Bromus tectorum* which shows local adaptations along an environmental gradient from arid steppe vegetation to subalpine forests in western North America (Rice & Mack 1991; Kollmann & Bañuelos, 2004).

Disturbance

Disturbance plays a prime role in invasion ecology. Disturbances can be defined as resource fluctuations which are discrete relative to the temporal scale of investigation (White & Pickett, 1985). There is an array of research works on varying ecosystems which studied the impacts of several disturbances on the success of invasive species and experimental studies have illustrated the role of disturbance in an invasion context. Disturbance may be in the form of fire as observed in South African fynbos which creates the space and provide an opportunity so that alien trees can establish themselves (Richardson & Cowling, 1992). Further, Fox & Fox (1986) concluded that "there is no invasion of natural communities without disturbance". Although anthropogenically modified disturbance regimes have, in particular, been implicated as invasion facilitators (Fox & Fox, 1986; Hobbs & Huenneke, 1992), invasions can occur under a natural disturbance regime (Richardson et al., 1992). Long term impacts of forest harvesting may result in invasion of exotic plant species (Marshall, 2000). Intensive grazing in forest ecosystems may exacerbate the alien problem particularly after fire (Keeley, 2004).

Hobbs (1989) showed that the presence of disturbed areas can enhance the establishment rate of invasive plants. Similarly, Bergelson et al. (1993) found that the area and spatial distribution of disturbed areas influenced invasive plant spread. Invasion case studies, through the use of correlative evidence have, like the experimental studies, also implicated disturbance. For example, DeFarrari & Naiman (1994) concluded from an alien plant survey that disturbance type and time since disturbance were the major factors influencing invasibility in Washington, USA. In another survey 90% of the alien species on Lord Howe Island, Australia, were associated with disturbed areas (Pickard, 1984). Similarly, the analysis by Crawley (1987) of floristic data of the British Isles revealed that aliens constituted more than 50% of the flora in highly disturbed areas, but less than 5% of the native woodland flora. MacDougall et al. (2006) found that low-stress environments were less invasible but appear to be more susceptible to invasion by species with strong competitive impacts.

Almost all this effort has been expended on observations of invasions in natural systems (de Waal et al., 1994; Pysek et al., 1995; Brock et al., 1997). There have been few attempts to study invasion experimentally, by manipulating either the characteristics of the invaded community or the identity of potential invaders (Peart & Foin, 1985; Robinson et al., 1995; Bastl et al., 1997; Tilman, 1997; Crawley et al., 1999; Knops et al., 1999; Levine, 2000; Thompson et al., 2001). Thompson et al. (2001) examined the roles of productivity and disturbance as major factors controlling the invasibility of plant communities, and simultaneously through field experiments identified the functional characteristics of successful invaders in response to different types of invasion opportunity. Field experiments established in 1990 comprised of seeds of 54 native species, not originally present at the site, were sown into fertility X disturbance matrix established in unproductive limestone grassland at the Buxton Climate Change Impacts Laboratory (BCCIL). Thompson et al. (2001) results were consistent with the hypothesis that invasions are promoted by an increase in the availability of resources, either through addition of extra resources or a reduction in their use by the resident vegetation. Volin et al. (2004) opined that invasion success of *Lygodium microphyllum* in disturbed as well as undisturbed greater Everglades ecosystem of southern Florida, USA may be attributed to the traits related to its reproduction, such as propagule pressure, and its ability to grow in a lowlight understorey environment. Maestre (2004) in his study in SE Spain along a disturbance gradient predicted that patch attributes are the major determinants of species richness and diversity in semiarid *Stipa tenacissima* steppes. Population and soil seed bank dynamics of *Nicotiana glauca* demonstrated that Disturbances in plant communities provide opportunities for weed germination, propagation, spread, and invasion (Florentine et al., 2006). Different invasive plants may evolve certain adaptable strategies in order to cope up with the disturbances e.g., accumulation of dormant but long-lived viable seed reserves serve as sources of regeneration of new *Prosopis juliflora* plants in the event of disturbance that might eliminate the aboveground stands (Shiferaw et al., 2004).

Leaf traits (specific leaf area (SLA), foliar nitrogen and phosphorus; N:P ratio) comparisons between natives and exotics at disturbed as well as undisturbed sites were significantly higher for exotics at disturbed sites compared with natives at undisturbed sites, with natives at disturbed sites being intermediate. Therefore, species with leaf traits enabling rapid growth will be successful invaders when introduced to novel environments (Leishman et al., 2007).

The role of road corridors in landscape fragmentation and disturbance, and as a reservoir of non-native plant species is scanty in literature (Angold, 1997; Forman & Alexander, 1998; Forman & Deblinger, 2000; Saunders et al., 2002; Gelbard & Belnap, 2003; Godefroid Koedam 2004). The altered disturbance regime in plant communities along corridor edges and vehicle traffic facilitate the spread and establishment of invasive non-native plant species (Hansen & Clevenger, 2005). Hansen & Clevenger (2005) compared the frequency of non-native plant species along highways and railways and the ability of these species to invade grasslands and dense forests along these corridors and emphasized the importance of minimizing the disturbance of adjacent plant communities along highways and railways during construction and maintenance, particularly in grassland habitats and in areas sensitive to additional fragmentation and habitat loss.

Chapter 8

Theories/Hypotheses/Factors for Invasion

Several theories have been proposed pertaining to ecological attributes or mechanisms responsible for invasion (**Figure 3**). Ultimately, it is unlikely that any single theory will be able to account for all differences in invasibility among all environments (Davis et al., 2000). A plant community becomes more susceptible to invasion whenever there is an increase in the amount of unused resources (Davis et al., 2000). In his recent article in Nature, Seastedt (2009) indirectly supported 'resource and enemy release hypothesis' (R-ERH), given by Blumenthal (2006). He opined that resource fluctuations and lack of enemies (fungal and viral pathogens) may act in concert, underpinning for invasion success. However, these two factors, instead of acting as drivers, merely act as passengers along for the invasion ride. Nevertheless, there exists a cascade of mechanisms behind the invasion success.

Enemy Release Hypothesis

Enemy release hypothesis reveals that if an organism introduced into a new region leaves behind its natural predators, competitors, and parasites, its chances of reproductive success increase. Competition for mutualistic interactors among exotic and native plant species provides another angle to the enemy release hypothesis (Keane & Crawley, 2002). Complete understanding of the role of enemy release in exotic plant invasions is prerequisite in order to prepare a comprehensive predictive model of exotic plant invasions (Keane & Crawley, 2002).

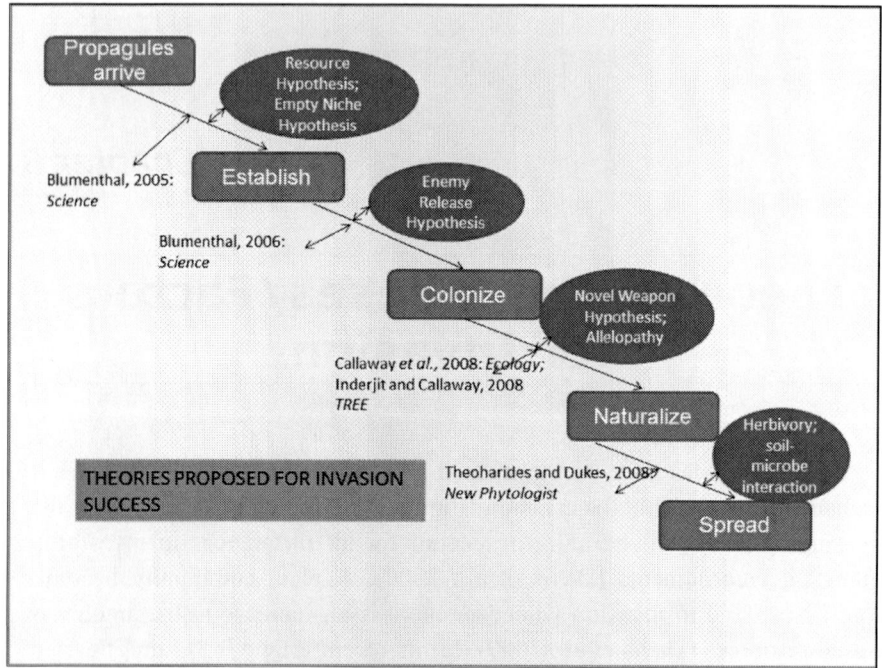

Figure 3. Salient hypothesis/theories applicable at varying invasion stages.

However, it is worth to mention that two of the most well known and best studied are the enemy release hypothesis (ERH) (Elton, 1958; Keane & Crawley, 2002; Blumenthal 2005, 2006) and the resource hypothesis (Davis et al., 2000; Blumenthal et al., 2003; Blumenthal, 2005; 2006) and both are ecologically interrelated (Blumenthal, 2005, 2006 Blumenthal (2005, 2006) emphasized that plant invasions is likely to involve not only multiple mechanisms of invasion (resource–enemy release hypothesis), but also understanding the conditions under which each mechanism tends to be important i.e., the effects of enemy release may be strongest for high-resource species. Increasing realization of aforesaid mechanisms may assist in the management of invasive species (Blumenthal, 2005, 2006). *Mikania micrantha* (mile a minute weed) is one of the top 10 worst weeds in the world (Table 1) mainly because of lack of natural enemies, a wide range of invasive habitats, and increased human disturbance associated with recent economic growth (Holm et al., 1977, Zhang et al., 2004).

Novel Weapon Hypothesis (NWH)

The novel weapons hypothesis raises the possibility of co-evolution among plants in different regions of the Earth, and that mixing species from different regions increases the chances of disrupting the ecological processes that lead to species coexistence and greater community diversity (Callaway & Ridenour, 2004; Vivanco et al., 2004). *Centaurea maculosa* (spotted knapweed), an invasive species in the western United States, displaces native plant species by exuding the phytotoxin (–)-catechin from its roots (Bais et al., 2003). Bais et al. (2003) demonstrated the allelopathic effects of *C. maculosa* by integrating ecological, physiological, biochemical, cellular, and genomic approaches and their results supported a "novel weapons hypothesis" (NWH) (Callaway & Aschehoug 2000, Callaway & Ridenour, 2004; Vivanco et al., 2004) for invasive success.

Resource (R)/Nutrients

One mechanism by which high resource availability might increase invasibility is by increasing the ability of non-native plants to compete with natives. Nutrients addition to soils (e.g., Wedin & Tilman, 1996; Bakker & Berendse, 1999; Kolb et al., 2002) as well as water (Milchunas & Lauenroth, 1995; White et al., 1997; Kolb et al., 2002) promotes invasion. Sharma & Raghubanshi (2009) studied impact of *Lantana camara* vegetative understory invasions on soil nitrogen (N) availability in forest ecosystems (Vindhyan forests, India) and observed alteration in litter inputs and chemistry beneath the lantana canopy positively and significantly altered soil N availability, N-mineralization, and total soil N. Another study (Osunkoya & Perrett, 2011)) demonstrated that under *Lantana* infested soil, moisture, pH, Ca, total and organic C, and total N were significantly elevated, while sodium, chloride, copper, iron, sulfur, and manganese, many of which can be toxic to plant growth if present in excess levels, were present at lower levels in soils compared to soils lacking *L. camara*. Likewise, garden and greenhouse experiments have shown that high nutrient or water availability can increase the ability of non-native plant species to compete with natives (Wedin & Tilman, 1993; Nernberg & Dale, 1997; Claassen & Marler, 1998; Kolb et al., 2002). In most cases, it has been demonstrated that the native species outperformed the alien under conditions of reduced light, nutrient or water availability (Daehler, 2003; Totland et al., 2005).

In mediterranean coastal dune ecosystem, long-term invasion by *Acacia longifolia* altered the soil properties with increased levels of organic C, total N and exchangeable cations resulting in higher microbial biomass, basal respiration, and b-glucosaminidase activity (Marchante et al., 2008). Further, Siemann et al. (2007) observed the impact of nutrient loading and extreme rainfall events on coastal tallgrass prairies found that it was more likely to be impacted by nutrient loading, in terms of invasion intensity. It has been demonstrated that microbial biomass C, N, and P all increased as the cover of *M. micrantha* increased, therefore, we can say that its invasion improved the soil attributes which inturn lead to its greater invasive success (Li et al., 2007).

Invasive plants may inhibit N-fixation and possibly lead to long-term declines in N inputs to soil (Wardle et al., 1994). The complex interactive effects of invasion on soil N-cycling are illustrated by grass invasion into submontane woodlands in Hawaii in which the grass increased net mineralization in the wet season due to changes in soil organic matter, but decreased net mineralization in the dry season due to decreases in soil moisture (Mack and D'Antonio, 2003; Bohlen 2006). Morghan & Seastedt (1999) in their study on non-native plants e.g., *Centaurea diffusa* and *Agropyron smithii* found that carbon amendment treatment (with sugar and sawdust) alone is not sufficient in order to sites deeply infested by invasive species.

Frequent small- scale disturbances, e.g., by burrowing animals, can create localized patches of unexploited resources, and thereby may facilitate invasions (Hobbs & Mooney 1985; Davis et al., 2000). Nitrogen addition in California serpentine grassland increased the invasion success of several alien grass species (Huenneke et al., 1990) and Harrison (1999) argued that the low invasibility of serpentine grasslands was due to low levels of soil nutrients. Maron & Connors (1996) concluded that invasions by exotic species in a California coastal prairie were facilitated by a native nitrogen-fixing shrub. Similar findings were found by Hobbs & Atkins (1988) who also found that disturbance combined with Eutrophication increased a community's invasibility. The combination of disturbance and eutrophication involves both a reduction in resource uptake by resident vegetation and an increase in gross resource supply.

Likewise, other studies (Burgess et al., 1991; Harrington, 1991; Hobbs & Mooney, 1991; Li & Wilson, 1998; Davis et al., 1999; Dukes & Mooney, 1999) have shown that in dry regions, increase of water supply increases the invasibility of vegetation, either as a direct effect of water supply or through improved access to mineral nutrients. Imposed drought conditions reduced the availability of soil water and hence decreased the invasibility of the same communities during the drought period (Davis et al., 1998, Davis et al., 1999). Some authors have predicted that the increase in atmospheric CO_2 will favour invasions by certain species by increasing soil water availability due to more efficient use of water by the resident plants (Idso, 1992; Johnson et al., 1993; Dukes & Mooney, 1999; Davis et al., 2000). This is an example of increased resource availability due to reduced uptake by the resident vegetation. Others have argued that invasions may be facilitated by increases in precipitation (Dukes & Mooney, 1999), an example of increased resource availability due to increased resource supply. Still others have argued that the global nitrogen In aquatic ecosystems, eutrophication resulting from anthropogenic activities is already facilitating invasions (Wedin & Tilman, 1996), another example of enhanced supply increasing resource availability.

Brooks (2003) showed that the negative impact of N addition and enhanced growth of nonnatives on native annuals in the Mojave desert only occurred in the year of highest abundance of annual plants, which in turn was regulated by winter rainfall.

Whether alien or native, species cannot maximize growth, reproduction and competitive ability across all environments, so the success of invasive species is habitat-dependent (Funk & Vitousek, 2007).). The interaction between habitat traits and intraspecific variation are found to be important when determining invasion success as experimented in case of invasive weed mugwort, *Artemisia vulgaris* (Barney et al., 2005). General assumption that invasive species colonize resource rich environment was contradicted by Funk and Vitousek (2007) who emphasized that invaders do colonize resource-poor environments and traits associated with resource conservation are widespread among species adapted to resource-poor environments employing resource conservation traits such as high resource-use efficiency.

Niche Opportunities/ Presence of Empty Niches

Community ecology theory can be used to understand plant invasions by applying recent niche concepts to alien species and the communities that they invade (Shea & Chesson, 2002). These ideas lead to the concept of 'niche opportunity', which defined conditions that promote invasions in terms of resources, natural enemies, the physical environment, interactions between these factors, and the manner in which they vary in time and space (Shea & Chesson, 2002). Recent niche theory clarifies the prediction that low niche opportunities/invasion resistance result from high species diversity. Conflicting empirical patterns of invasion resistance are potentially explained by co-varying external factors. These various ideas derived from community ecology provide a predictive framework for invasion ecology (Shea & Chesson, 2002).

Invasion success also relies on suitability of niche dimension (Badano & Pugnaire, 2004). In this context, *Agave* species in Spain can be cited which showed higher rhizome and bulbil production, and higher establishment rates by agaves in sandy soils than in clay soils (Badano & Pugnaire, 2004). This suggested that sandy soils provide an opportunity which releases the clonal reproduction of *Agave*. Furthermore, the effects of agaves on the physiological performance and reproduction of native species were negative, positive or neutral, depending on the size and rooting depth of neighbours (Badano & Pugnaire, 2004).

Allelopathic Advantage against Resident Species (AARS)

A logical extension of the NWH is the hypothesis that populations of invaders in the invaded region should evolve greater concentrations of allelopathic, defense or antibiotic biochemicals than populations of the species in their native range (Inderjit et al., 2008). This hypothesis has been called the 'allelopathic advantage against resident species' (AARS). If invaders possess allelochemical weapons that provide greater competitive advantages in their new habitats than in their original ranges, then selection may act directly on those traits (Callaway & Ridenour, 2004). The fundamental prediction of AARS is that invasive populations will be more allelopathic, or better biochemically defended, than source populations (Callaway & Ridenour, 2004; Inderjit et al. 2005, Inderjit et al. 2008). Allelopathic effects not only derive from the release of phytotoxins from roots, but also from biochemicals present in leaves that leach during rainfall or release chemicals after senescing and falling to the ground (Inderjit & Keating 1999; Inderjit et al., 2008). Plant invasion is a huge and complex area that encompasses many aspects in addition to the potential contribution of Allelochemicals (Field et al., 2006). Allelopathy is one such mechanism that has been implicated in the success of at least some of the best known plant invaders in the world (Weir et al., 2004; Allaie et al., 2006), including *Eltrygia repens* (Weston et al., 1987; Korhammer & Haslinger, 1994; Allaie et al., 2006), *Bromus tectorum* (Rice 1974; Allaie et al., 2006), *Circium arvense* (Stachon & Zimdahl, 1980; Allaie et al., 2006), *Cyperus rotundus* (Agarwal et al., 2002), *Eichhornia crassipes* (Gopal & Sharma, 1981; Allaie et al., 2006), *Lantana camara* (Saxena, 2000; Allaie et al., 2006), *Parthenium hysterophorus* (Kanchan & Jayachandra, 1980; Kohli & Batish, 1994; Pandey, 1994; Tefera, 2002; Singh et al., 2003; Allaie et al., 2006) and *Prosopis* spp. (Goel et al., 1989). The most credible evidence for the involvement of allelopathy in plant invasion has come from the study of Bais et al. (2003) and Callaway & Aschehoug (2000) on *Centaurea* species in North America. Several biochemical and physiological processes of the susceptible species are affected by the allelochemicals released by such invasive species in their environment (Weir et al., 2004). Invasive species like *Eupatorium adenophorum*, dominant in some part of North East India is shown to exhibit allelopathic impact (Tripathi et al., 1981). In riparian habitats Reinhart et al. (2005) demonstrated that *Acer platanoides* trees suppress most native species, including the regeneration of the natural canopy dominants, but facilitate conspecifics in their understories.

Several arguments on allelopathy lead to the hypothesis that plant species that normally coexist should evolve resistance to each others' toxins, but not to the toxins of species with which they do not coexist i.e., invasive species (Fitter, 2003). The observation by Bais et al. (2003) on *C. maculosa* provided strong evidence for the differential susceptibility of target plants to a phytotoxin ((−)-catechin), and supports the discovery of an ecologically important allelopathic interaction. The toxin promotes production of reactive oxygen species in the roots of susceptible plants, which activates a Ca^{2+} signaling cascade that initiates extensive changes in gene expression and death of the root system (Bais et al., 2003). Leaf leachate of *Anthemis cotula* inhibits seed germination of the commonly cultivated field crops in Kashmir valley (Allaie et al., 2006).

Biochemical Basis of Invasion

Biochemical basis is an extension or explanation for AARS as mentioned earlier. The invading species produces a phytotoxin, catechin, which induces oxidative stress in many native plants and often thereby eliminates them entirely from the local ecological community (Inderjit et al., 2008). Further, the findings of Inderjit et al. (2008) highlighted the role for the biochemical potential of the plant as an important determinant of invasive success besides earlier existing enemy release hypothesis (Kennedy et al., 2002) pointing that invasiveness is mainly due to enhanced resource competition after escape from natural enemies. *Cyperus rotundus* (nutgrass) is the world's worst invasive weed through tubers since its extract inhibits acetylcholinesterase activity from animal and plants as well as inhibits germination and seedling growth in wheat and tomato (Sharma & Gupta, 2007). Apart from case study on *Centaurea* sp., root exudates from *Acroptilon repens* (Russian knapweed) were found to be phytotoxic and the phytotoxin in the exudate was identified as 7,8-benzoflavone (a-naphthoflavone) (Stermitz et al., 2003).

Many invasive weeds, however, eventually encounter their native, coevolved enemies in areas of introduction (Zangerl & Berenbaum, 2005; Zangerl et al., 2008). Examination of herbarium specimens of an invasive phytotoxic European weed, *Pastinaca sativa*, through 152 years reveals phytochemical shifts coincident in time with the accidental introduction of a major herbivore, the parsnip webworm, *Depressaria pastinacella* (Zangerl & Berenbaum, 2005; Zangerl et al., 2008). In the aforesaid reassociation with a coevolved specialist in invaded area profoundly altered the selection regime, favouring trait remixing and rapid chemical changes in parsnip populations, as predicted by the geographic mosaic theory (Zangerl et al., 2008). That uninfested populations of New Zealand parsnips contain higher amounts of octyl acetate, a floral volatile used by webworms for orientation, suggests that plants that escape from specialized enemies may also experience selection to increase kairomones, as well as to reduce allomones (Zangerl et al., 2008).

Evolution of Increased Competitive Ability (EICA)

Among the more specific hypotheses considered was the evolution of increased competitive ability (EICA) of plants in the absence of specific herbivores (Bais et al., 2003). The EICA hypothesis predicts that once an organism escapes its natural enemies, it no longer needs the defences it had evolved against them. Dana Blumenthal of the U.S.D.A. marked the "very compelling examples and evidence that EICA can occur," in meeting of

Ecological Society of America (ESA, 2004). Absence of herbivores may result in selection for the loss of costly herbivore-resistance traits, which are expected to show a trade-off with vigour or competitive ability (the evolution of increased competitive ability, or EICA, hypothesis). Statstny et al. (2005) demonstrated through his garden experiment that increased competitive ability of invasive plants may be associated with changes in resistance as well as tolerance to herbivory, and both types of anti-herbivore defence may need to be examined simultaneously to advance our understanding of invasiveness.

The better performance of *Solidago gigantean* (Asteraceae), an invasive species in Europe, as compared to North American range may be the result of changed selection pressures, as implied by the EICA hypothesis (Jakobs et al., 2004).

In enemy-free space, resources previously used for herbivore defence become dispensable and can be reallocated to growth and reproduction. Such changes can result in the evolution of highly competitive, but less well defended genotypes (Blossey & Nötzold, 1995). Increased competitive ability can emerge not only from the growth and defence tradeoff, but also from weak co-adaptation between native and invasive species (Callaway, 2002). Callaway (2002) showed that allelopathic root interaction creates a competitive advantage for *Centaurea maculosa* in invaded North American communities, but not among species of its native range. In enemy-free space, resources previously used for herbivore defence become dispensable and can be reallocated to growth and reproduction. Such changes can result in the evolution of highly competitive, but less well defended genotypes.

Propagule Pressure

Propagule pressure is extremely important factor, required initially for invasion success. Baker (1955, 1967, 1974) proposed that self-compatible plants, particularly those capable of autonomous self pollination, are most likely to be successful as colonists on account of their ability to establish populations from a single propagule after long-distance dispersal. This idea was termed 'Baker's Law' by Stebbins (1957). Controlled pollination experiments carried out on 17 invasive alien plant species in South Africa revealed that 100% were either self-compatible or apomictic, and that 72% of these were capable of autonomous self pollination. The distribution of breeding systems among these invasive aliens is thus strongly skewed towards uniparental reproduction (Rambuda & Johnson, 2004). Thus Baker's rule, which has generally been considered for short-lived herbaceous plants, may also apply to invasive shrubs and trees (Rambuda & Johnson, 2004). Insect pollination between colonizers could moderate bottleneck effects when colonization begins with a few or scattered individuals (Regal, 1977). Interplay of seed and pollen dispersal systems also play a pivotal role during initial establishment stage and further it is intimately linked with ecology as well as evolution of plant groups (Regal, 1977). The study of impact of soil parameters on invasion revealed that site history and propagule pressure may be more important in determining exotic species success than soil characteristics alone, in this vegetation community (Hill et al., 2005).

Pollination success in diverse habitats e.g., in the case of *Lantana camara, Ligustrum robustum, Mimosa pigra* through profuse nectar and prolonged flower production (Ghazoul, 2002) aid in their invasion success. *Solanum mauritianum* recovers rapidly after clearing, and previously heavily invaded cleared sites due to both prolific resprouting recovery from cut stumps and through seedling emergence from the dense soil seed bank in the more open post-clearing environment (Witkowski & Garner, 2008). Alien plants *Ligustrum robustum, Tibouchina herbacea, Lantana camara* and *Mimosa pigra* might gain the double advantage of appropriating pollinators at the expense of natives whilst ameliorating seed predation simply by virtue of relatively higher seed set (Ghazoul, 2002). At community level, the overwhelming effects of ecological factors spatially covarying with diversity, such as propagule supply, make the most diverse communities most likely to be invaded (Levine, 2000). In *Prosopis juliflora,* production of many, small and hard seeds capable of surviving passage through the digestive system of animals, entering into the soil to form soil seed banks and remaining viable until favourable conditions for germination and seedling establishment appear (Shiferaw et al., 2004)

In contrast with the aforesaid findings, Bellingham et al. (2004) investigated the association of plant species invasiveness with seedling relative growth rate and survival, among 33 naturalized woody plant species in four families (Fabaceae, Mimosaceae, Pinaceae, Rosaceae), however, failed to find a consistent theoretical positive relationship and hypothesized that simple life history trait such as seedling relative growth rate provides a general explanation for patterns of plant invasion success in disturbed habitats.

Competition (Plant-Plant Interaction)

Plant-plant interaction should also be taken into account in invasion ecology particularly in the context of competition (Callaway, 1995; Brooker, 2006). Invasive species evolve in response to their interactions with natives as well as in response to the new abiotic environment and concomitantly alter the evolutionary pathway of native species by competitive exclusion, niche displacement, hybridization, introgression, predation, and ultimately extinction (Mooney & Cleland, 2001).

Competitive plant–plant interactions commonly play a central role in invasion ecology (Brooker, 2006). Obligate mutualistic relationships among species are ubiquitous and central to ecological function and the maintenance of biodiversity (Palmer et al., 2008). The extensive review of Traveset & Richardson (2006) concluded that invasive species frequently cause profound disruptions to plant reproductive mutualisms.

Generally, invasive species are not dominant competitors in their natural systems, but competitively eradicate their new neighbours as revealed in case of *Centaurea diffusa*, a noxious weed in N. America, observed to be more aggressively competingon grass species from N. America than on closely related grass species from communities to which *Centaurea* is native (Callaway & Aschehoug, 2000).

Brooker (2006) in his critical review correlated plant-plant-interaction with global environmental change and emphasized that competition plays a central role in mediating the impacts of atmospheric nitrogen deposition, increased atmospheric carbon dioxide concentrations, climate change and invasive nonnative species. For example, in the native dry forest ecosystems of Hawaii, the dense roots and shoots of invading grass species negatively affect nutrient and water acquisition and germination of native woodland species (D'Antonio & Vitousek, 1992; Cabin et al., 2002; Brooker, 2006), whilst in Californian coastal chaparral communities the invasive *Carpobrotus edulis* reduces soil water availability to native shrubs, negatively affecting their growth and reproduction (D'Antonio & Mahall, 1991; Brooker 2006). In both these cases the type of interaction is one that the native species will have experienced before, i.e., diffuse competition for resources such as water or nutrients (Brooker, 2006). Positive relationship between fitness and population size (density) in small populations i.e., Allee effect (Allee, 1931), is a mechanism by which plant– plant interactions might have a selective impact. At low densities, reduced seed set and recruitment can occur as a consequence of pollen limitation (Antonovics & Levin, 1980; Davis et al., 2004; Brooker

2006). Flexibility in behavior, and mutualistic interactions, can aid in the success of invaders in their new environment (Mooney & Cleland, 2001).

Vila & Wiener (2004) reviewed pair-wise experiments between invading and native plant species in order to test the hypothesis that invasive plants often appear to be more competitive than native species. Most importantly it has been suggested that the influence of an invading species on total plant community biomass is an important clue in understanding the role of competition in a plant invasion (Vila & Wiener, 2004).

Role of Aboveground and Belowground Communities

In terrestrial ecosystems, soil microbes are important regulators of plant diversity as well as affecting invasion, especially in nutrient poor ecosystems where plant symbionts are responsible for the acquisition of limiting nutrients (van der Heijden et al., 2008). The 9th biennial meeting of the Soil Ecology Society held in Palm Springs, CA in May 2003, addressed the theme of "Invasive species and soil ecology" (Bohlen, 2006).

In invasion ecology, aboveground and belowground communities can be powerful mutual drivers, with both positive and negative feedbacks (Grime, 2001; Wardley et al., 2004). However, belowground invasions may be equally widespread. Exploring links between above and belowground communities illuminates the broader ecological implications of species invasions (Wardle, 2002).

Root-associated organisms and their consumers influence plants more directly, and they also influence the quality, direction, and flow of energy and nutrients between plants and decomposers. Exploration of the interface between population- and ecosystem-level ecology is an area attracting much attention (Wardley et al., 2004) and requires explicit consideration of the aboveground and belowground subsystems and their interactions. Invasive plants can also alter ecological interactions in the rhizosphere leading to important but poorly understood consequences for microbial dynamics, nutrient uptake and competitive interactions in the plant community (Bohlen, 2006).

Soil biota in some invaded ecosystems may promote 'exotic' invasion, and plant–soil feedback processes are also important (Callaway et al., 2004). Two of the most economically and ecologically damaging invasive plants on

North American rangelands are diffuse knapweed (*Centaurea diffusa* Lam.) and spotted knapweed (*Centaurea maculosa* auct. Non Lam.) (Lacey et al., 1989; Roche, 1994; Sheley et al., 1998). Presently, these two Eurasian knapweeds are widely distributed across North America (Sheley et al., 1998; USDA NRCS, 2002). Relative benefit of native soil communities to two native plants and two knapweeds i.e., diffuse (*Centaurea diffusa* Lam.) and spotted knapweed (*Centaurea maculosa* auct. non Lam.) and the growth of these plants in soil from knapweed infestations and from adjacent native rangelands were investigated in North America (Meiman et al., 2006). Meiman et al. (2006) observed that native soil community appeared to be more beneficial to spotted knapweed than to the other plants studied, including diffuse knapweed. Therefore, it appears that two closely related knapweeds have very different interactions with soil biota and perhaps different strategies for invasion (Meiman et al., 2006). Callaway et al. (2004) reported that soil microbes from the home range of the invasive exotic plant *Centaurea maculosa* L. have stronger inhibitory effects on its growth than soil microbes from where the weed has invaded in North America. In invaded soils, *Centaurea* cultivates soil biota with increasingly positive effects on itself, which may contribute to the success of this exotic species in North America (Callaway et al., 2004). Kornissa & Caraco (2005) applied the physical theory for nucleation of spatial systems to a lattice-based model of competition between plant species, a resident and an invader, and the analysis reaches conclusions that differ qualitatively from the standard ecological theories.

Callaway et al. (2008) found that one of North America's most aggressive invaders of undisturbed forest understories, *Alliaria petiolata* (garlic mustard) which inhibits mycorrhizal fungal mutualists of North American native plants, has far stronger inhibitory effects on mycorrhizas in invaded North American soils (attributed to specific flavonoid fractions in *A. petiolata* extracts) than on mycorrhizas in European soils where *A. petiolata* is native.

Mangla et al. (2008) demonstrated a new pathway/mechanism on experimenting with microbial role (*Fusarium semitectum*) in rhizosphere soils of *Chromolaena odorata* and on native species which indicated that the impacts of this severe tropical weed are due to the exacerbation of biotic interactions among native plants and native soil biota rather than just enemy release or novel interaction hypothesis.

One of the most apparent and dramatic examples of belowground invaders is the invasion of northern forest by non-native earthworm species, a subject that has received much attention in recent years (Bohlen et al., 2004a, b, Bohlen 2006). In the case of earthworms, much of their effect occurs because of their role as ecosystem engineers capable of substantially changing the physical and chemical characteristics of the soil environment, with consequences for the entire soil food web, nutrient distribution, and even vertebrate and understory plant communities (Bohlen, 2006). Species-site characteristics actually determine the impact of the invasive weed plants on the soil microfauna rather than invasive/ native species in isolation (Yeates & Williams, 2001).

Microcosm investigations indicated that the composition of the arbuscular mycorrhizal fungi (AMF) community belowground can influence the structure of the plant community aboveground, and may play a role in facilitating or repelling invasion (Stampe & Daehler, 2003).

In plant invasion also study of this interrelationship (above ground/below ground-invasion) is imperative (Wardle et al., 2004). A combined aboveground-belowground approach to community and ecosystem ecology is enhancing our understanding of the regulation and functional significance of biodiversity and of the environmental impacts of human-induced global change phenomena (Wardley et al., 2004).

Soil microbes have profound negative and beneficial effects on plants through pathogenic effects, root–fungus mutualisms and by driving the nutrient cycles on which plants depend (Callaway et al., 2004). Callaway et al. (2004) demonstrated that soil microbes from the home range of the invasive exotic plant *Centaurea maculosa* L. have stronger inhibitory effects on its growth than soil microbes from where the weed has invaded in North America. Therefore, *Centaurea* and soil microbes participate in different plant–soil feedback processes at home compared with outside *Centaurea*'s home range. In native European soils, *Centaurea* cultivates soil biota with increasingly negative effects on the weed's growth, possibly leading to its control. But in soils from North America, *Centaurea* cultivates soil biota with increasingly positive effects on itself, which may contribute to the success of this exotic species in North America (Callaway et al., 2004).

Root-derived natural products play an important role pertaining to interactions between plants and soilborne organisms, by serving as signals for initiation of symbioses with rhizobia and mycorrhizal fungi (Field et al., 2006). They may also contribute to competitiveness of invasive plant species by inhibiting the growth of neighbouring plants through the mechanism of Allelopathy (Field et al., 2006). It has been demonstrated through various researches that the ability to produce allelopathic chemicals may contribute to success of invasive plants (Whittaker & Feeney 1971; Rice, 1984; Williamson,

1990; Callaway & Aschehoug, 2000; Inderjit & Duke, 2003; Callaway et al., 2005; Field et al., 2006; Inderjit et al. 2008).

Root-derived natural products play an important role pertaining to interactions between plants and soilborne organisms, by serving as signals for initiation of symbioses with rhizobia and mycorrhizal fungi (Field et al., 2006). They may also contribute to competitiveness of invasive plant species by inhibiting the growth of neighbouring plants through the mechanism of Allelopathy (Field et al., 2006). It has been demonstrated through various researches that the ability to produce allelopathic chemicals may contribute to success of invasive plants (Whittaker & Feeney, 1971; Rice, 1984; Williamson, 1990; Callaway & Aschehoug, 2000; Inderjit & Duke, 2003; Callaway et al., 2005; Field et al., 2006; Inderjit et al., 2008).

Insurance Hypothesis

Increasing domination of ecosystems by humans is steadily transforming them into *depauperate systems* (Loreau et al., 2001). A major future challenge is to determine how biodiversity dynamics, ecosystem processes, and abiotic factors interact. The insurance hypothesis (Yachi & Loreau, 1999) propose that biodiversity provides an "insurance" or a buffer, against environmental fluctuations, because different species respond differently to these fluctuations, leading to more predictable aggregate community or ecosystem properties (Yachi & Loreau, 1999; Loreau et al., 2001). Microbial microcosm experiments show less variability in ecosystem processes in communities with greater species richness, perhaps because every species has a slightly different response to its physical and biotic environment (Naeem & Li, 1997).

Diversity Resistance Hypothesis

Biological invasions are a pervasive and costly environmental problem that has been the focus of intense management and research activities over the past half century (Kennedy et al., 2002). The diversity resistance hypothesis, which argues that diverse communities are highly competitive and readily resist invasion, is supported by both theory and experimental studies conducted at small spatial scales (Elton 1958; Crawley, 1987; Case, 1990; McGrady-Steed et al., 1997; Tilman, 1997, 1999; Levine & D'Antonio, 1999;

Knops et al., 1999; Levine, 2000; Naeem et al., 2000; Symstad, 2000; Dukes, 2001; Kennedy et al., 2002).

Herbivore Pressure

Intensive herbivory by ungulates can enhance exotic plant invasion, establishment (de Villalobos et al., 2011), and spread because: (1) many exotic plants are adapted to ground disturbances such as those caused by ungulate feeding, trampling, and movements; (2) many exotic plants are adapted for easy transport from one area to another by ungulates via endozoochory and epizoochory; (3) many exotic plants are not palatable or are of low palatability to ungulates, and consequently, their survival is favored as ungulates reduce or eliminate palatable, native plants (Hobbs et al., 1996; Augustine & McNaughton, 1998; Riggs et al., 2000; Kie & Lehmkuhl, 2001; Riggs et al., 2005; Vavra et al., 2007). In order to study how the predators impact the succession of vegetation, we derive invasion conditions under which a plant species can invade into an environment in which another plant species co-exists with a herbivore population with or without a predator population (Feng et al. 2011). Horse grazing is reported to be responsible for invasion success of *Pinus halepensis* as mentioned in Table 1 (de Villalobos et al., 2011).

Herbivore-resistance traits of dominant plant species and impacts of "keystone" animal species cascade through the system to affect many organisms and ecosystem processes (Brown et al., 2001).

Herbivore pressure may be an important attribute facilitating the invasion process as demonstrated with the spatial pattern of colonization by the *prairie* lupin, *Lupinus lepidus*, which is governed by herbivore pressure. The plants are eaten by the leaf-tying larvae (caterpillars) of several lepidopteran species, and there is evidence for thresholds in the parameter ranges of plant spatial extent and timing of initial colonization that predict whether the herbivores can halt the invasion (Chin, 2005). As well as providing fresh insight into the dynamics of successional systems, these findings are relevant to the control of invasive plants because they suggest the possibility of developing protocols for the most effective timing and spatial deployment of herbivorous control agents (Chin, 2005). Results of Clay et al. (2005) have broad implications for understanding the success of invasive species. They mentioned that plants invading novel habitats may frequently suffer less damage from pests and parasites than native species. Further, in their experiment, the relative biomass of infected tall fescue was enhanced by herbivores, suggesting that this grass may be better able to invade novel habitats with high levels of herbivore pressure. Moreover, their results confirmed the important role of mammalian herbivores in shaping the composition and dynamics of plant communities (Clay et al., 2005). Parker et al. (2006) performed meta-analysis of 63 manipulative field studies including more than 100 exotic plant species which revealed that native herbivores provide biotic resistance to plant invasions, but the widespread replacement of native with exotic herbivores eliminates this ecosystem service, facilitates plant invasions, and triggers an invasional "meltdown."

Ecological factors like fire often increases the abundance and diversity of exotics (Hughes et al., 1991; Milberg & Lamont, 1995; D'Antonio et al., 2000), resulting in a positive feedback enhancing the dominance of the exotic grasses and more intense fires (D'Antonio & Vitousek, 1992) (Figure 4), as demonstrated in forest of Amazon basin. Invasive African grasses in the Amazon are having highly flammable litter leading to forest fire and hence conversion of forests to grass land/savannah. Land use change leads to less carbon sequestration and hence contributing to global climate change (Figure 4).

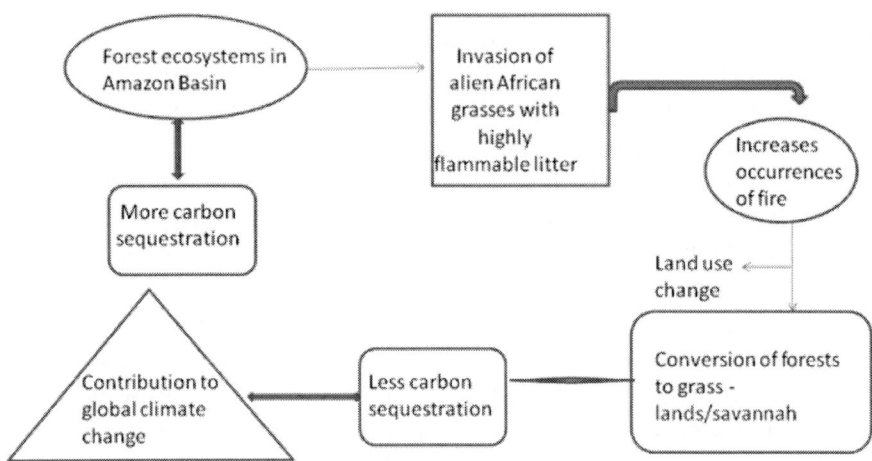

Figure 4. Interface of invasion with land use and global climate change [Modified after D'Antonio and Vitousek (1992); Mack et al. (2000)].

In addition to aforesaid ecological attributes/ hypotheses/ theories, Raffaeli (2004) mentioned certain attributes which may be equally applicable to biodiversity loss emanating from invasive species. Attributes such as body size and its related characteristics such as home range and tolerance to stress, together with differences in species richness between trophic levels, will determine the impact on ecosystems of different biodiversity loss scenarios (Raffaelli, 2004).

Evolutionary Aspect

Despite the increasing biological and economic impacts of invasive species, little is known about the evolutionary mechanisms that favour geographic range expansion and evolution of invasiveness in introduced species (Sax, 2005; Lavergne & Molofsky, 2007). Eco-evolutionary consequences of climate change and role of the emerging synthetic discipline of evolutionary community ecology is predicted to have a profound impact on future biodiversity (Lavergne et al., 2010). Lavergne & Molofsky (2007) focused on the genetic and evolutionary aspects of invasive species through study on wetland grass *Phalaris arundinacea* L. in North America demonstrating rapid selection of genotypes with higher vegetative colonization ability and phenotypic plasticity and hence tendency to evolve in response to changing climate. Further, multiple immigration events of *Phalaris arundinacea* L., thus trigger future adaptation and geographic spread of a species population by preventing genetic bottlenecks and generating genetic novelties through recombination (Lavergne and Molofsky, 2007). Using a phylogenetic supertree of all grass species in California, Strauss et al. (2006) showed that highly invasive grass species are, on average, significantly less related to native grasses than are introduced but noninvasive grasses.

Charles Elton (1958), who stated, ''we must make no mistake: we are seeing one of the great historical convulsions in the world's fauna and flora.'' Elton certainly had no doubts of the magnitude of the invasive species issue. More recently, Geerat Vermeij (1996) remarked specifically about the evolutionary consequences of this convulsion, ''. . . if newcomers arrive from far away as the result of large-scale alterations in geography or climate, the change in selective regime and the evolutionary responses to this change could be dramatic.'' (See Mooney &Cleland. 2001).

Evolutionary aspect of invasive species also drew attention of researchers (Lee 2002). Further, workshop on the Evolutionary Perspective of Biological Invasions in Terrestrial Ecosystems was held in Halle, Germany from 30 September to 3 October 2002 which discussed evolutionary issues pertaining to invasive species at a length. To mention a few outcomes of conference, Bernd Blossey from Cornell University, Ithaca, USA demonstrated that

Lythrum salicaria, introduced individuals exhibited greater biomass than did their ancestral genotypes and were less resistant to, or tolerant of herbivores. William Rogers (Rice University, Houston, TX, USA) demonstrated increased competitive ability of invasive populations of the tree *Sapium sebiferum* and elimination of this advantage when herbivores from the native range are abundant. Klaas Vrieling (University of Leiden, the Netherlands) presented a special twist of the EICA concept, showing that invasive populations of Asteraceae *Senecio jacobaea* had reduced adaptation to a specialist herbivore, but greater defence against generalist insect herbivores (See Hänfling & Kollmann, 2002). Jes Pedersen, University of Copenhagen, Denmark demonstrated that Argentine ant *Linepithema humile*, share an extraordinary social structure called 'unicoloniality', where individuals from physically separated nests mix freely, and form supercolonies. Therefore, changes in life-history traits are also observed in invasive animals.

Recent studies (Tsutsui et al., 2000; Ellstrand & Schierenbeck 2000, Filchak et al., 2000; Krieger. & Ross, 2002; Lee, 2002) suggested that the invasion success of many species might depend more heavily on their ability to respond to natural selection than on broad physiological tolerance or plasticity (Lee, 2002). Lee (2002) in his extensive review on evolutionary genetics of invasive species emphasized the utility of exploring genomic characteristics of invasive species, such as genes, gene complexes, and epistatic interactions that promote invasive behaviour. Such information could yield insights into the relationship between genetic architecture and rate of evolution, and evolutionary versus ecological factors which confer invasion success. Ellstrand & Schierenbeck (2000) in their concise review demonstrated that hybridization between species or between disparate source populations may serve as a stimulus for the evolution of invasiveness. Eurasian *Tamarix* plant species (potent novel hybrids) have spread rapidly to dominate over 600,000 riparian and wetland hectares in US (Gaskin & Schall 2002).

The extreme elevation gradients in the Hawaiian Islands provide specific opportunities for comparative studies on the ecology and evolution of temperate invaders while also creating a unique field environment for understanding interactions between temperate and tropical species (Daehler, 2005) and Daehler (2005) found that number of naturalized species declined exponentially with increasing altitude, however, in contrast, the proportion of species of European or Eurasian origin appeared to increase linearly with elevation, from 38% among all species occurring above 1200m to 53% above 2000m and 90% above 3000m.

Genetic Diversity/Hybridization

As we know that biodiversity is the sum total of all biotic variation from the level of genes to ecosystems (Purvis & Hector, 2000), genetic diversity is also an important factor particularly in relation to host-pathogen co-evolution and prospecting of disease resistant genes (Allen et al., 2004). Because elucidating how allelic diversity within plant genes that function to detect pathogens (resistance genes) counteracts changing structures of pathogen genes required for host invasion is critical to our understanding of the dynamics of natural plant populations (Allen et al., 2004). Genetic studies are decoding the language plants and microbes use to negotiate the symbioses and genes from both plants and microbes contribute to symbiosis (Marx, 2004). The two partners engage in a complex molecular conversation that allows the microbes to infect the plant cells and then entice the cells to undergo the developmental changes necessary for establishing the symbioses (Marx, 2004).

Willis et al. (2000) tried to test the hypothesis that increases size of certain invasive weeds is genetical rather than environmental and found that actually it is a plastic response to novel environment. Threats to biodiversity e.g., habitat fragmentation prevent sufficient dispersal of natives whereas long term dispersal in case of exotics tend to maintain their genetic connectivity and hence invasion success (Trakhtenbrot, 2005). Moreover, gene manipulation in different disciplines require utmost precaution as it has generated concern over the risk of producing new invasive species or exacerbating current weed problems (Parker & Kareiva, 1996).

Genetic modifications, through traditional breeding or genetic engineering, of crop or other species can potentially create changes that enhance an organism's ability to become an invasive species (Wolfenbarger & Phifer, 2000). Although genetic engineering transfers only short sequences of DNA relative to a plant's entire genome, the resulting phenotype, which includes the transgenic trait and possibly accompanying changes in traits, can produce an organism novel to the existing network of ecological relationships. Potential ecological impacts through invasiveness depend on existing opportunities for unintended establishment, persistence, and gene flow of an introduced organism; each of these, in turn, depends on various components of survival and reproduction of an organism or its hybrids. Few introduced organisms become invasive, yet an issue for the management of all introduced organisms, including GEOs, is how to identify those modifications that may lead to or augment invasive characteristics (Wolfenbarger & Phifer, 2000).

The transition from colonist to invader is especially enigmatic for self-incompatible species, which must find a mate to reproduce (Elam et al., 2007). Elam et al. (2007) conducted a field experiment to test whether the Allee effect affects the maternal fitness of a self-incompatible invasive species, wild radish (*Raphanus sativus*) and observed that both population size and the level of genetic relatedness among individuals influence maternal reproductive success. Even polyploids e.g., hexaploids *Carthamus creticus* and *C. turkestanicus* are noxious weeds with wide but non-overlapping Mediterranean distributions (Vilatersana et al., 2007).

Demographic studies i.e., germination and seedling survivorship characteristics of hybrids between native and alien species of dandelion (*Taraxacum*) were studied and observed that *T. platycarpum* (4X) have the advantage over *T. officinale*, whose seedlings could not survive under high temperatures (Hoya et al., 2004).

Pollen Sharing

It is a form of interaction between plants. The physiological characteristics of plants from an invasion front have been shown to differ from those in the source population in such a way as to overcome the negative consequences of these Allee effects, i.e., they have a greater degree of self-compatability, indicating that changes in the strength of interactions may have acted as a selective force (Davis, 2005; Brooker, 2006). According to Baker's Rule, plant species capable of uniparental reproduction are more likely to be successful colonists than are self-incompatible or dioecious species (Rambuda & Johnson, 2004).

Chapter 9

Interelationship of Invasion with Climate Change

Invasive species can act synergistically with other elements of global change, including land-use change (Vitousek et al., 1996; Hobbs, 2000; Theoharides. & Dukes, 2007), climate change (Dukes & Mooney, 1999; Simberloff, 2000; Kriticos et al., 2003; Theoharides & Dukes, 2007; Ghermandi et al., 2010; Crossman et al., 2011), increased concentrations of atmospheric carbon dioxide and nitrogen deposition (Dukes & Mooney, 1999; Dukes, 2002; Weltzin et al., 2003; Theoharides. & Dukes, 2007) (Figures 1,4,5,6). Exotic plant invasions and its linkage with changing climate and other human caused disturbances impact the intensity of biological invasions is extremely important (Siemann et al., 2007). Fire, drought and plant invasions are likely to increase the effects of any climate change induced damage (Foster, 2001).

Benning et al. (2002) illustrated that interactions of climate change with land-use change and biological invasions exacerbate the direct effects of climate change substantially. The effects of climate, atmospheric changes and invasion are likely to grow in importance, particularly because they influence communities and their component species over their entire ranges (Zavaleta et al., 2003).

Climate change and species invasion are not actually independent and have an inextricable linkage as demonstrated by taking insect herbivores as model (Ward and Masters, 2007). Climate change will affect population attributes e.g., propagule pressure and the communities into which invaders will arrive and also likely to increase niche-availability in the future thus exacerbating the problem of invasive species (Sakai et al. 2001, Ward & Masters, 2007) (Figure 5). Moreover, unprecedented global warming may affect plant species physiology, phenology, distribution, tree species abundance, distribution, forest community characteristics (e.g., species richness and diversity) and biome distributions (Sakai et al., 2001; Hughes, 2000; Hansen et al., 2001; Xu et al., 2007; Crossman et al., 2011) (Figure 6). CO_2 induced increases in optimum temperature could substantially reduce forest landscape change caused by global warming (Xu et al., 2007). Sherry et al. (2007) predicted that phenological divergence originating from climate warming may create a potential reproductive niche for heat-tolerant species in mid-summer, which may be conducive to invasion by nonnatives (Figure 6). Alteration in climate is supposed to affect the interactions between soil biota and its structural processes (Young et al., 1998) which also may have indirect implication on climate changes.

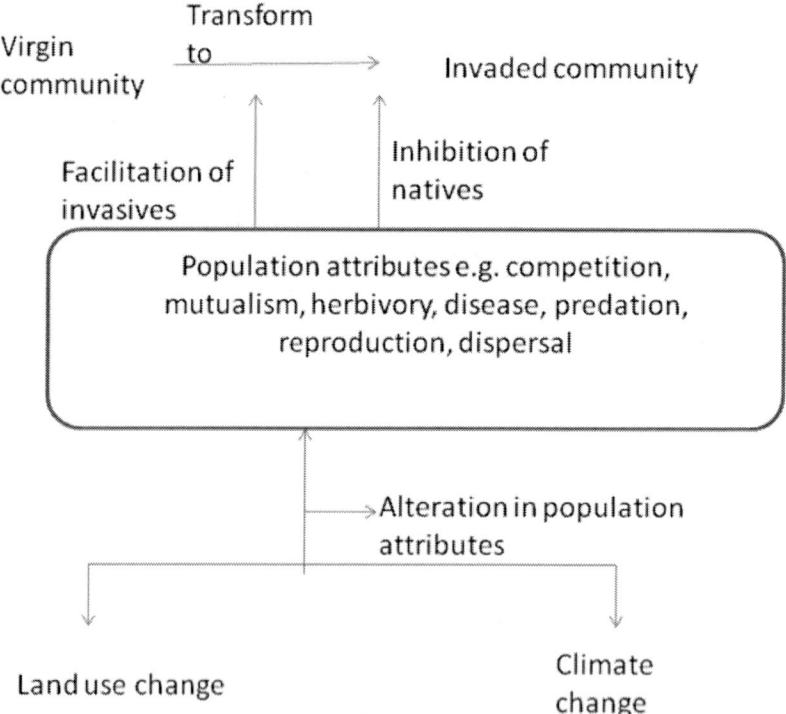

Figure 5. Effect of other anthropogenic perturbations on population attributes and hence success of invasive plants over native ones.

Several studies demonstrated that invasion may act in concert with climate or land use change. In an integrated study on the islands of Kauai and Hawaii, Benning et al. (2002) showed that anthropogenic climate change is likely to combine with past land-use changes and biological invasions (particularly vectors of human diseases) to drive several of the remaining species to extinction.

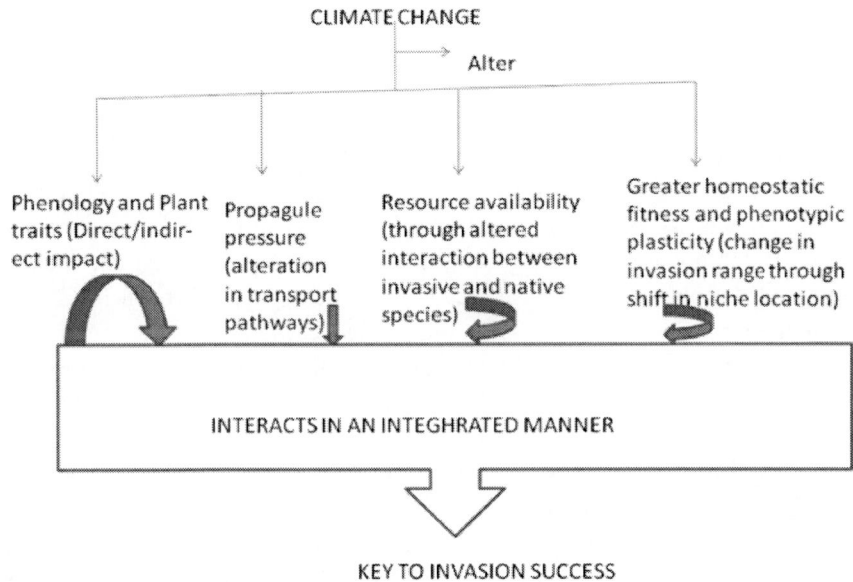

Figure 6. Interlinked attributes responsible for invasion success [Adapted from Ward and Masters (2007)].

New evidence suggested that many invasive species share traits that will allow them to capitalize on the various elements of global change (Dukes & Mooney, 1999). Increases in the prevalence of some of these biological invaders would alter basic ecosystem properties in ways that feed back to affect many components of global change (Dukes & Mooney, 1999). Many invasive plants have been shown to respond positively to elevated CO_2 when grown individually or in monoculture. Examples of these are species that have invaded North America, such as cheatgrass (*Bromus tectorum*), kudzu (*Pueraria lobata*) and Japanese honeysuckle (*Lonicera japonica* (Smith et al., 1987; Sasek & Strain, 1987, 1991; Taylor & Potyin, 1997; Dukes & Mooney, 1999).

Climate change or global warming may facilitate a shift to dominance by non-native species, accelerating the homogenization of the global biota (Stachowicz et al., 2002). Using three most abundant introduced species of ascidian, Stachowicz et al. (2002) compared their long-term record with interannual variation in water temperature to assess the likely effect of climate change on their success and spread. Their data suggested that the greatest effects of climate change on biotic communities may be due to changing maximum and minimum temperatures rather than annual means. Rouget et al. (2004) made an assessment of the climatic correlates of distribution of 71 important invasive alien plants, and an analysis of the implications of these findings for future invasions in different vegetation types existing in South Africa, Lesotho and Swaziland over the next few decades.

Climate induced invasion may exhibit more severe impact on biodiversity hotspot although its extent may vary. The lower montane cloud forests of the eastern Andes are a biodiversity hot spot and among the most threatened habitats on the planet (Bush et al., 2004). In this region, Bosh et al. (2004) though their continuous 48,000-year-long paleoecological record found that the warming rate of about 1°C per millennium during the Pleistocene-Holocene transition was an order of magnitude less than the projected changes for the 21st century.

Plant species respond less predictably to CO_2 enrichment when they are grown in diverse communities (Taylor & Potvin, 1997; Duke & Mooney, 1999). Consequently, the response of *Toxicodendron* to global environmental change, particularly the current increase in global atmospheric carbon dioxide (CO_2) concentrations, bears consequences for human health on a panoptic scale (Mohan et al., 2006).

Apart from climate change, extreme climatic events e.g., heavy or uneven rainfall may also affect community composition and dynamics (Thibault & Brown, 2007). Thus, Long-term monitoring of community dynamics provides unique opportunities to observe extreme climatic events and to document their ecological impacts (Thibault & Brown, 2007).

Chapter 10

Management Options for Invasive Species

There is an urgent need for the prioritization of collaborative research on invasion ecology in order to promote its sustainable management (Donlan et al., 2003). Some studies gave impetus to extensive monitoring in order to assess the impacts of invasives while some relatively limited data sets may allow us to draw reliable inferences for adaptive management in the context of ecological restoration and rehabilitation (Fleishman et al., 2005). Ecosystem based approach and inventory of priority areas is extremely urgent for sustainable invasive plant species management (Goodall & Naude, 1998).

Effective and sustainable management of the impacts of invasives on native diversity may be achieved through a better compromise between research and practical conservation needs and well directed focus based research (Byers et al., 2001). In order to formulate a sustainable management plan for invasive plants, it is necessary to develop a synoptic view of the dynamic processes involved in the invasion process (Kriticos et al., 2003). Driesche and Driesche (2000) made it clear in their book that the roots of invasion problems, and possible solutions, are a complex mixture of biology, sociology, economics and history. Foxcroft et al. (2011) reviewed conceptual framework of plant invasion ecology and emphasized the inclusion of three contributing processes for explicit understanding i.e., (1) the characteristics of the introduced species, (2) system context, within which the invasion takes place, and (3) the features of the receiving habitat.

Also, SCOPE, the Scientific Committee on Problems of the Environment, chartered symposia in North America, Australia, and Britain to deal with three main questions: *What factors determine whether a species will become an invader or not? What site properties determine whether an ecological system will accept or repel invaders? And, given answers to those questions, how can we best manage ecological systems to avoid problems?* Answers to this set of questions are important not only with respect to management issues but also for understanding how natural systems work (See Case, 1987). Therefore, first question emphasized the traits of 'invaders', the third addressed management and the second emphasized researches to identify site properties that determine whether an environment would be succeptible to invasion or not (Davis, 2009). In order to curb the future economic and environmental impacts of invasive exotic species, we need to understand the mechanisms behind exotic invasions. Kean & Crawley (2002) emphasized that in relation to ERH, there is a serious need for experiments involving exclusion of natural enemies in invaded plant communities. With a clearer understanding of the role of enemy release in exotic plant invasions, we can begin to build a comprehensive predictive model of exotic plant invasions (Kean & Crawley, 2002). Strauss et al. (2006) advocated that a match between characteristics of the invader and those of members of the existing native community may be essential to understanding invasiveness and hence suggesting the management approaches. Now, we will discuss the different management prerequisites and options for the success of biological invasions. In forthcoming section, different means which may be implemented for the management of invasive species are discussed. However, some sections may be applicable to one group of invasive plants and not others.

Understanding of Mechanism

In this regard, understanding of the mechanisms behind invasion may be helpful in the management of plant invasions. In conjunction with Resource-Enemy Release Hypotheses, plant invaders follow other interrelated mechanisms in complementarities, at different invasion stages i.e., transport, colonization, establishment, and landscape spread (Figure 2). 'Evolution of increased competitive ability' (EICA) through 'novel weapons (NW) hypothesis' as demonstrated in case of root exudates of *Centaurea diffusa* (the Eurasian forb, diffuse knapweed; NW: 8-hydroxyquinoline) and *C. maculosa* (spotted knapweed; NW: (±)-catechin) in North America is mentioned in previous sections. NW hypothesis has also been termed as the 'allelopathic advantage against resident species' (AARS) Similarly, 'an empty niche (EN) hypothesis is extremely relevant in species poor or fragmented landscapes, providing sites for colonization. Therefore, establishing an inextricable linkage between abiotic and biotic thresholds is imperative particularly in depauperate systems (Theoharides, K.A. & Dukes, 2007). Considering multiple traits, mechanisms and native-alien interactions in terms of competition (-) and facilitation (+) is pertinent, instead of drawing an unjustified distinction between native and aliens. Furthermore, synergistic response of native-alien interaction to global environmental changes (e.g., fragmentation, climate change, nitrogen deposition) is an area that is ripe for investigation (Figures 5, 6).

Fruit traits e.g., fruit size, the presence of an inedible peel, defensive chemistry, crop size and phenology may be certain traits for frugivores involved in spreading the invasive plants (Liddy, 1985; Buckley et al., 2006) which is further promoted by landscape fragmentation and an integrated approach may assist in framing of sophisticated mechanistic seed dispersal models for surveying, eradicating and managing plant invasions. Seedling growth of natives is also affected by biological invasions which are tremendously important for restoring the native diversity as demonstrated in the case of riparian habitats (Sweeney & Czapka, 2004).

Proper management of roadside plant communities is prerequisite in order to prevent their spread into the forest interiors (Gelbard & Belnap, 2003) as demonstrated in a study on exotic plant species in the southern boreal forest of Saskatchewan (Sumners & Archibold, 2007).

It has been observed that the height and ground cover of *Ageratina riparia* belonging to family Asteraceae was positively associated with light availability beneath the rainforest canopy and negatively associated with native forest leaf litter biomass, thus providing an ecological implication for management (Zancola et al., 2000). As demonstrated in the case of *Lantana camara*, forest gap/canopy openness plays a major role in invasive spread therefore canopy intactness may be the one of the prime management strategy which is rather difficult to maintain (Totland et al., 2005). Further, through different manipulation/disturbance experiments on *L. camara* it has been recommended that shading by intact canopies is an effective barrier against its successful invasion and hence, is the most appropriate strategy for managing invasion (Gentle & Duggin, 1997; Duggin & Gentle, 1998). It has been observed that in *L. camara* seed set was strongly correlated with honeybee (*Apis mellifera*) abundance; therefore, management of honeybee populations may provide a powerful tool for its cost-effective control (Goulson & Derwent, 2004).

Geographical features and landscape ecology also affect the invasion process and they deals with the mechanism through which spatial pattern e.g., habitat fragmentation and others act on invasion process in an integrated way (With et al., 2001). In order to reduce the negative consequences of fragmentation landscape corridors are being created in which habitat patches connected by corridors retain more native plant species than do isolated patches, and prevent invasion by exotic species (Damschen et al., 2006). Null and logistic regression models suggested that species richness is a poor indicator of invasive resistance and there should be ecological study of invasion attributes at landscape scale (Higgins et al., 1999).

Mason & French (2007) emphasized that management regimes relying on intensive as well as extensive strategies had adverse impacts on species richness or composition thus management regimes must be considered a form of disturbance. Current invasive plant control methods can, however, adversely affect indigenous fauna (Gosper, 2004).

Exploitation of Benefits

As mentioned earlier, plant invasion is now considered to be one of the 'big five' environmental issues of public concern, however, not all invasive plants are deleterious. Some introduced species, like corn (*Zea mays* L.), wheat (*Triticum* spp.), rice (*Oryza sativa* L.), plantation forests, domestic chicken (*Gallus* spp.), cattle (*Bos taurus*), and others, are beneficial and provide more than 98% of the world's food supply (Pimentel et al., 2001; Sharma et al., 2005). It has been a matter of argument that invasive species differ from native species through the characteristics of rapid growth, fewer pathogen burden and fewer herbivores (Seastedt, 2009; Blumenthal et al., 2009). Comparisons were also made at global datasets between agricultural and natural weeds pertaining to their taxonomic status and invasive strategies which revealed that plants with amongst the highest risk of becoming natural area invaders worldwide include species that are primarily aquatic or semi-aquatic, grasses, nitrogen-fixers, climbers, and clonal trees (Daehler, 1998).

Sometimes invasive species play a rare positive role in establishment of native species particularly in degraded ecosystems because they come first at these sites and create conditions for native species. *Cinnamomum camphora* (Camphor laurel), an exotic, fleshy fruited tree re-growth provides habitat for rainforest birds and creates conditions suitable for the regeneration of native rainforest plants on abandoned farmland (Neilan et al., 2006). Also, fleshy-fruited invasive plants provide food that supports indigenous frugivore populations which duly assist in restoration of indigenous plants in an effective way. *Lantana* has got many benefits also e.g., lantana addition improves soil hydraulic properties to the benefit of the wheat crop in a rice–wheat cropping sequence (Bhushan & Sharma, 2005). Further, it may act on other invasives e.g., the growth of the aquatic weed *Eichhornia crassipes* and the alga *Microcystis aeruginosa* may be inhibited by fallen leaves of *Lantana camara* (Kong et al., 2006). However, contrastingly Mascaro et al. (2008) investigated 46 sites and found that native species represent a very small and probably decreasing share of understory plant diversity in novel Hawaiian forests, and therefore found no evidence to support the use of exotic tree species in restoration activities in which the goal is to promote native plant regeneration. *Prosopis juliflora* also has great ecological (seed and impermeable seed coat has potential to survive in the wider range of environment) and socio-economical importance (due to production of woody biomass, may be useful in social forestry) (Shiferaw et al., 2004).

Exotic species may also act as ecosystem engineers causing physical state changes (Crooks, 2002). Although the consequences are difficult to predict, however, invasion offers an opportunity to predict what will be impact upon their integration into systems. Crooks (2002) suggested that invasive ecosystem engineers that increase habitat complexity or heterogeneity tend to cause abundances and/or species richness to rise, while those that decrease complexity tend to have the reverse effect.

Fleshy-fruited invasive plants provide food that supports indigenous frugivore populations (Gosper & Vivian-Smith, 2006) and Gosper & Vivian-Smith (2006) using *Lantana camara* as a target species suggested that using the fruit characteristics of the invasive plant may assist to select replacement indigenous plants that are functionally similar from the perspective of frugivores.

Invasive plants also perform important ecological functions (Westman, 1990; Buckley et al., 2006; Gosper & Vivian-Smith, 2006). In California, for example, some butterflies have expanded their geographical ranges or extended their flight season as a consequence of being able to feed on alien plants (Graves & Shapiro, 2003; Gosper & Vivian-Smith, 2006). In northern New South Wales, Australia, several indigenous frugivores appear to rely on the fruits of Camphor Laurel (*Cinnamomum camphora* (L.) Nees) as their principal food over part of the year (Date et al., 1996; Gosper & Vivian-Smith, 2006). In both cases, the alien and/or invasive plants may have helped buffer these fauna populations from the broad-scale destruction of their natural habitats (Date et al., 1996; Graves & Shapiro 2003). Relative to methods of achieving control, the ecological consequences of control programs have received little research attention (Williams & West, 2000).

Eradication

Earlier, eradication was increasingly being realized as an easy approach to manually remove the invasive plants. However, eradication alone might not allow ecosystems to recover, because some invaders change the condition of the habitat so as to render it unsuitable for native species (Zavaleta et al., 2001). For instance, in sites from the Middle East to the western USA, high soil salinity is caused by the invasive ice plant *Mesembryanthemum crystallinum*, and tamarisk *Tamarix* spp., which makes it difficult for salt-sensitive native species to re-establish (El-Ghareeb et al. 1991, Zavaleta et al., 2001). In these cases, eradication must be followed by additional site restoration (Zavaleta et al., 2001).

Moreover, invasive plant species removal in isolation can also result in unexpected changes to other ecosystem components. These secondary effects will become more likely as numbers of interacting invaders increase in ecosystems, and as exotics in late stages of invasion eliminate native species and replace their functional roles (Zavaleta et al., 2001). Food web and functional role frameworks can be used to identify ecological conditions that forecast the potential for unwanted secondary impacts. Integration of eradication into a holistic process of assessment and restoration will help safeguard against accidental, adverse effects on native ecosystems (Zavaleta et al., 2001).

The best solution is to prevent the introduction of exotic organisms but, once introduced, eradication might be feasible. The potential ecological and social ramifications of eradication projects make them controversial; however, these programs provide unique opportunities for experimental ecological studies (Myers et al., 2000).

Biocontrol Agents

Due to the high costs and risk of failure of biocontrol agents (Markin et al., 1992; Briese, 1993; Kriticos et al., 1999), it is worthwhile considering factors that might indicate the effectiveness of a potential agent. McConnachie et al. (2003) advocated an understating on financial mechanism to address biological control of invasive species in a sustainable manner. Biological control as a tool to fight invasions has got certain negative implications which can be achieved through host-specific agents and also number of introduced biological agents should be minimum with maximum potential for control (Pearson & Callaway, 2003).

Sheppard et al. (2006) checked the feasibility of biological control on top 20 invasive species of Europe in terms of ecological, social and economic perspective and found that the method is effective in chemicals management. Biological control can be better verified to reduce the impacts among native plants could be better predicted by knowledge of characteristic chemical profiles of secondary compounds (Jordon-Thaden & Louda, 2003). Recently, the flowerhead weevil (*Rhinocyllus conicus* Fro¨l.), introduced from Europe into North America to control exotic thistles (*Carduus* spp.), has become invasive and have limitation in controlling certain *Cirsium* spp (Jordon-Thaden & Louda, 2003). *Chromolaenu odorata* (L.) King & Robinson (= *Euputorium odorutum* L.), a perennial, semi-woody, herbaceous scrambling shrub of neotropical origin, is an invasive weed of agriculture, forestry and conservation in many parts of the world (Goodall, J.M., Erasmus, 1996).

Bryophyllum delagoense (Crassulaceae), is a major weed in Queensland, Australia which can be biologically controlled with *Alcidodes sedi* (Col.: Curculionidae) (Witta et al. 2004). In India, *Puccinia spegazzinii*, was released in India in 2005 for the classical biological control of the invasive alien weed, *Mikania micrantha*, a notorious weed of North-Eastern Himalayas (Ellison et al., 2007). Baars (2003) studied the lepidopteran association with *L. camara* and found that additional agents are required to supplement its biological control in South Africa. At continent level The European Biological Control Laboratory (EBCL) is developing biological control technologies to be used to suppress invading weeds e.g., *Centaurea* spp., *Arundo donax*, *Vincetoxicum* spp., *Isatis tinctoria*, *Taeniatherum* sp. and *Dipsacus* sp. (Jones and Sforza, 2007). The genetic uniformity of invasive *A. donax*, revealed through using Sequence Related Amplification Polymorphism (SRAP) and transposable element (TE)-based molecular markers generated scope for the classical biological control of the species (Ahmad et al., 2008).

Many pros and cons are also associated with the impact of biological control on natives; therefore, process of ecological risk assessment established by the United States' Environmental Protection Agency may be followed to improve assessment of the risks of proposed biological control agents

(Andersen et al., 2005). Biological control agents should be integrated with traditional methods in order to have sustainable invasive weeds management strategy as observed in the case of leafy spurge (*Euphorbia esula*) (Lym, 2005).

Searching the Biochemichal Tools

A few native species, such as *Gaillardia grandiflora*, are able to resist knapweed invasion, and several of these species, including *Lupinus sericeus*, facilitate the resistance of native grasses to the invader. *Lupinus* secretes oxalate from its root tissues in response to catechin exposure. By blocking reactive oxygen species, oxalate affords protection to neighboring vulnerable plants against the toxic effects of catechin. These results suggest strategies for controlling a serious invader and also provide insight into the multiplicity of facilitation mechanisms involved as plant communities develop.

Remote Sensing and Geospatial Technologies

Early detection of invasive plants through remote sensing particularly through hyperspectral imagery for mapping invasive species may be fruitful in devising control strategies (Hamada et al., 2007). Remote sensing has been used to map the distribution of some biological invaders (Lass et al., 2002; Underwood et al., 2003; Asner & Vitousek, 2005), but invasions often must be far advanced before they can be detected by conventional remote sensing that relies on canopy emergence and/or dominance by the invader (Hunt et al., 2004; Asner & Vitousek, 2005).

Asner & Vitousek (2005) used airborne imaging spectroscopy and photon transport modeling to determine how biological invasion altered the chemistry of forest canopies across a Hawaiian montane rain forest landscape. Asner & Vitousek (2005) demonstrated that *Myrica faya* (nitrogen fixing plant) doubled canopy nitrogen concentrations and water content as it replaced native forest, whereas the understory herb *Hedychium gardnerianum* reduced nitrogen concentrations in the forest overstory and substantially increased aboveground water content.

With increasing availability of fine-scale climatic data linked with geographic information systems (GIS), there is significant interest in building and testing more sophisticated models that forecast an invader's eventual range, based on its distribution elsewhere (Sutherst & Maywald, 1991; Sutherst et al., 2000; Peterson & Vieglais, 2001; Peterson, 2003; Daehler, 2005). Relative Risk Model (RRM) was proposed in order to assess the regional impact on invasive species which quantitatively ranks sources of stressors and habitats by using Geographic Information Systems (GIS) to analyze spatial datasets to determine risk at the regional level (Colnar et al., 2007).

Despite the advent of technologies like remote sensing, it has limitations in assessing the threats imposed from invasive species (Brooks et al., 2006), and need more advancement pertaining to monitoring as well as management of invasive species.

Modelling Techniques

Mathematical modelling has provided a very useful tool for understanding complex dynamics of ecosystems/community as well as population attributes (Ghermandi et al., 2010; Feng et al., 2011; Järemo & Bengtsson, 2011; Lemke et al., 2011; Myster, 2012). Several modelling techniques also assist in assessment mapping and distribution of invasive plants (Lemke et al., 2011). Robertson et al. (2004) devised a new predictive modelling technique called the fuzzy envelope model (FEM) in order to predict potential distributions of organisms using presence-only locality records and a set of environmental predictor variables. FEM was applied to predicting the potential distribution of three invasive alien plant species (*Lantana camara* L., *Ricinus communis* L. and *Solanum mauritianum* Scop.), and three native cicada species (*Capicada decora* Germar, *Platypleura deusta* Thun. and *P. capensis* L.) in South Africa, Lesotho and Swaziland (Robertson et al., 2004). One model attempted to reveal certain predictors of invasiveness i.e., 'fast growth rate', 'native latitudinal range', and 'growth form' (Herron, 2007). Trethowan et al. (2011) did ecological niche modelling to predict which areas in southern Africa are likely to be suitable for pompom weed and its two potential biological control agents mentioned in **Table 1** and models indicated that pompom weed is likely to spread across a greater region of southern Africa, however, the Savannah and Grassland biomes being at greatest risk of invasion.

For *Fallopia japonica* (Japanese knotweed), which causes substantial economic and environmental damage in United Kingdom (UK), a 3D correlated random walk model of the development of the subterranean rhizome network for a single stand of *F. japonica* was constructed (Smith et al., 2007). Aforesaid case study was an exception to the existing trait that clonally/vegetatively producing plants are not very successful invaders.

Goslee et al. (2001) suggested individual plant-based simulation model (ECOTONE) to evaluate the importance of allelopathy and soil texture to the invasion of semiarid grasslands by the non-native perennial C_3 forb *Acroptilon repens* and found that Allelopathic interactions were an important component of the invasion dynamics of this perennial invasive weed.

Integrative models e.g., SPAnDX can be used as a decision-support tool in integrated weed management, and to explore the sensitivity of cultural management practices to climate change as already demonstrated in case of *Acacia nilotica* (Kriticos et al., 2003). Crossman et al. (2011) used the dispersal attribute to develop a management index for identifying invasive plant threat under climate change. Gravuer et al. (2008) analyzed the invasion of *Trifolium* (true clover) species into New Zealand, assessing a range of human, biogeographic, and biological influences at three key invasion stages: introduction, naturalization, and spread and used sparse principal component analysis (SPCA) to define suites of related attributes and aggregated boosted trees to model relationships with invasion outcomes. Their (Gravuer et al., 2008) results highlighted key roles that humans can play in facilitating plant invasion via two pathways, firstly, commercial introduction leading to widespread planting and concomitant naturalization and spread and secondly unintentional introduction and spread of species associated with human activities, such as seed contaminants.

Sa´ nchez-Floresa et al. (2008) used a scaled-down modelling approach, based on field data and high-spatial resolution imagery, to assess the predictive skill of combined genetic algorithm rule set-production (GARP) and change vector analysis (CVA) models in order to confirm the hypothesis that highly dynamic desert environments are unstable, thus more vulnerable to invasion by exotic plant species than stable landscapes.

Modelling current and potential invasive distribution lays groundwork for the application of management strategies that prevent or mitigate adverse effects on native ecosystems (Van Devender et al., 1997). *Predictive models look* for functional relationships between environmental conditions and the occurrence of invasives based on their particular ecological requirements or niches (Grinnell, 1917). These relationships remain, however, difficult or impossible to characterize completely because they arise from complex ecological and biotic interactions and are influenced by the genotypic plasticity of the invasives that favors adaptation (Williamson & Fitter, 1996). After decades of modeling, ecologists have yet to identify the natural 'rules' that govern processes of invasion and, thus, have strong predictive value (Bright 1998). Predictive modeling of invasives is thus subject to uncertainties because of the many factors intervening in their establishment (Rouget et al., 2004). Complexities of the plant's development and structure *Arundo donax* L. (a perennial reed), an invasive weed of riparian systems in North America, were attempted to be investigated using L-system modelling using structural model, L-DONAX (Thornby et al., 2007).

Several recent studies have demonstrated the importance of using statistical techniques and modelling that incorporate the issue of space on some level when assessing species distributions (Powell, 1990; Higgins & Richardson, 1996; Higgins et al., 1999; Jetz & Rahbek, 2002; Dark 2004). For instance, Higgins et al. (1999) found that using logistic regression models and incorporating a spatial element in the model output improved their ability to predict the distribution and impact of invasive plant species on native plant species. Jetz & Rahbek (2002) found that spatially explicit regression models improved their ability to predict the characteristics that determine bird species richness in sub-Saharan Africa. Dark (2004) emphasized the critical importance of testing for spatial autocorrelation in spatial pattern studies and using Spatial Autoregressive Model (*SAR*) models in his comparative study on the distribution of invasive alien plants ($n= 78$) and noninvasive alien plants ($n= 1097$) which revealed that distribution of both categories of alien plants was similar with the exception of a higher concentration of invasive alien plants in the North Coast bioregion.

Chapter 11

Discussions and Future Prospects

It is a general assumption that in sustainability we take care of future outcomes in terms of economy and environment. Aforesaid statement also implies when we attempt to manage the environmental perturbations in order to maintain environmental sustainability. In previous section, we covered different mechanisms responsible for biological invasions. However, forecasting of invasion as well as its impact may be better strategy for their management. In ecological sciences we also attempt to manage the threats arising out of the problems well in advance or ahead of time e.g., invasive species in our case as it may aggravate in future endeavour. van Wilgena (2007) used data on the current and potential future distribution of 56 invasive alien plant species to estimate their impact on four services (surface water runoff, groundwater recharge, livestock production and biodiversity) in five terrestrial biomes and demonstrated that the current impacts of invasive alien plants are relatively low (with the exception of those on surface water runoff), the future impacts could be very high. The main strength of the meeting on Strategic issues for European biodiversity research, held in Agropolis, Montpellier, France from 4 to 6 December 2000 was the consideration given to ecological science as a basis to understanding the problem of invasive species (Scott, 2001). A Partial analysis, combining knowledge of the rate of spread, seedbank longevity, costs of control and techniques of economic analysis, can assist in making a good decision (Cacho et al., 2007).

Individual researchers, journals, and ecological societies all can take specific steps to increase the useful exchange of ideas and information among research areas. Promoting rapid and more effective communication among diverse researchers may reduce the proliferation of narrow theories, concepts, and terminologies associated with particular research areas (Davis et al., 2005). In this way, we can expedite our understanding of the ecological mechanisms and consequences associated with plant invasions.

Moreover, plant invasion ecology can benefit greatly from further and better comparisons at regional and global scales. Comparisons among areas with similar climates have the advantage that some features of the abiotic environment are within a narrower range of variation, enabling the researcher to focus on the effects of propagule pressure, microenvironmental differences and, more importantly, the biotic environment in the invasion process (Pauchard et al., 2004).

Anthropogenic global perturbations threaten species and the ecosystem services inextricably linked with society. Macroecology is a recent research program that aims to develop quantitative predictions of the abundance and distribution of species, usually over broad areas i.e., landscapes (Kerr et al., 2007) and may find an application in the area of invasion ecology.

Apart of distribution and abundance, more researches should focus on the biochemical basis of invasion. According to Field et al. (2006) we are certainly making progress in exploring the tip of the metabolic iceberg in plants, and in learning about how this chemical repertoire is deployed for plant defence.

Lewine (1987) mentioned that the ecological attributes of 'good invader' plays most important role during initial establishment phase however, we should also focus on reasons for failure of invasive species instead of overemphasizing the mechanisms of their success in a particular environment in order to have better management strategy. Therefore "an understanding of the effects of invading species at the ecosystem level is important for conservation biology as well as for basic ecology" (Lewine, 1987). Despite much investigation, it has proven difficult to identify traits that consistently predict invasiveness (Alpert et al., 2000). This may be largely because different traits favour invasiveness in different habitats. It has proven easier to identify types of habitats that are relatively invasible, such as islands and riverbanks (Alpert et al., 2000).

Before the assessment of impact in the form of extinction of plant biodiversity, utmost care should be taken. Gurevitch & Padilla (2004) demonstrated that the impact of invasion on species extinction is overemphasized and stated that only 6% of the taxa are threatened with extinction as a result of invasion, however, aforesaid assumption has been challenged as extinction is complex process and many factors may act in conjunction.

There are scanty researches on invasion problems in tropical forests which are of extreme ecological relevance. Despite the proved threats from invasive plant diversity to native biodiversity in ecosystems (Vitousek et al., 1997; Mooney, 1999; Totland et al., 2005), invasive species have rarely been considered as a significant threat to the diversity of tropical forests (Fine, 2002; Totland et al., 2005), and while invasions have been the subject of intensive ecological research during the last two decades, this research has largely ignored tropical forests (Drake et al. 1989, Williamson 1996, Totland et al. 2005). Sheil (2001) described the grimy situation of conservation and biodiversity monitoring in tropics exacerbated by financial constraints in developing countries. One recently completed project TREE which visualized the rate of deforestation of global humid tropical forests through advanced remote sensing tools revealed that Between 1990 and 1997, 5.8 ±1.4 million hectares of humid tropical forest were lost each year, with a further 2.3 ±0.7 million hectares of forest visibly degraded (Achard et al., 2002). One reason for this neglect of tropical forests is the assumption that their high species diversity makes them naturally resistant to invasion (Holdgate, 1986), however as revealed by the diversity-invasibility conflict, tropical forests may not be overlooked.

Application of molecular techniques may further be advanced in order to study the ecological attributes responsible for invasion in totality. Random Amplified Polymorphic DNAs (RAPDs) were performed on three populations of *Juniperus ashei* i.e., ancestral, one growing on limestone substrate and blackland soil of Texas (Adams et al., 1998) which revealed that new mutations or allelic combinations has enabled *Juniperus ashei* to invade diverse habitats in United States (Adams et al., 1998). Therefore, molecular characterization of invasive species may also assist in unravelling the mechanism of their success. Godfree et al. (2006) developed a general quantitative framework for predicting changes in realized niche size and intrinsic population growth rate after introgression of disease resistance genes into wild host populations. They (Godfree et al., 2006) then applied this framework to a model host–pathogen system targeted by genetically modified and conventionally bred disease-resistant host lines (*Trifolium repens* lines expressing resistance to *Clover yellow vein potyvirus*) and showed that, under a range of ecologically realistic conditions, the introduction of novel pathogen resistance genes into host populations can pose a quantifiable risk to associated nontarget native plant communities.

Modelling of the complex interrelated ecological factors may provide an insight to the plant invasion. Since most mathematical models use relationships between parameters and variables as an abstraction of some ecological processes, a conceptual understanding of invasion processes (Figure 1-6) is imperative (Higgins and Richardson, 1996).

Future research should be in line with establishing the link between allelopathy as evidenced in case of *Centaurea* species and other invasive taxa e.g., *Sorghum, Acroptilon, Senecio, Zoysia, Tribulus, Cyperus, Lonicera Alliaria* etc. and should focus on the relative and combined roles of different ecological processes that influence overall plant biochemistry in realistic field conditions in native and invaded regions (Hierro & Callaway 2003, Inderjit et al. 2008). Ultimately, different invaders can succeed through different strategies and circumstances; however, creating an overall ecological, biochemical and genomics invasive plant model should help to understand how plants become invasive and to design ways to manage this growing global problem (Inderjit et al., 2008).

The evolutionary genetics of invasive species has been relatively unexplored, but could offer insights into mechanisms of invasions (Lee, 2002).

As discussed in the beginning of the present review that plant invasions may impact both in the terms of economy and ecology, however, it will be better to have an integrated assessment approach. Taylor and Irwin (2004) proposed a hypothesis that links ecology and economics to provide a causal framework for the distribution of exotic plants in the United States. They incorporated two approaches i.e., population-only model, which included direct effects of human population and ecological factors as predictors of exotics while the population-economic model including the direct effects of economic and ecological factors and the indirect effects of human population as predictors of exotics. Their hypothesis revealed suggested the incorporation of economics matter for resolving the exotic-species problem because the underlying causes, and insights into vegetation dynamics are likely to come from an expanded synthesis of evolution, population, community, and ecosystem ecology; from additional comparative, observational, and experimental studies; and from theory that links these together (Rees et al., 2001).

Policy Framework

According to the Convention on Biological Diversity (CBD) national strategies have to be developed to deal with biological invasions (Born et al., 2005). CBD exhorts the Contracting Parties to "prevent the introduction, control or eradicate those alien species which threaten ecosystems, habitats or species" (Glowka et al., 1994, Williamson, 1999). In this regard, SCOPE, the Scientific Committee on Problems of the Environment, has launched GISP (Dean, 1998), the Global Invasive Species Programme, with support from the United Nations (GEF, the Global Environmental Facility, and UNEP, the United Nations Environment Programme), IUCN (now World Conservation Union) and others (Williamson, 1999).

"Trends in invasive alien species" is one of only two indicators of threat to biodiversity that form part of the Convention on Biological Diversity's (CBD) framework for monitoring progress toward its "2010 target" (i.e., the commitment to achieve by 2010 a significant reduction in the current rate of biodiversity loss). To date, however, there is no fully developed indicator for invasive alien species that combines trends, derived from a standard set of methods, across species groups, ecosystems, and regions (Mcgeoch et al., 2006).

The best method by which progress toward the 2010 target is to be monitored remains a matter of active debate (Balmford et al., 2005; Pereira & Cooper, 2006). "Trends in invasive alien species" is one such headline indicator that is not yet ready to be implemented and that requires further development (Mace et al., 2005, but see UNEP, 2005). For example, an evaluation of this indicator is not included in the special issue of *Philosophical Transactions of the Royal Society of London B* (2005, volume 360) that addresses several other of the CBD headline indicators and general principles relevant to all of them. Nonetheless, the CBD has proposed "numbers and cost of alien invasions" as a possible measure under this headline indicator (UNEP, 2004*a*). The relevant CBD framework goal is to "control threats from invasive alien species" and the two targets are to (1) "control pathways for major potential alien invasive species" and to (2) "have management plans in place for major alien species that threaten ecosystems, habitats or species" (UNEP, 2004*a*, 2004*b*, 2005). An indicator of trends in invasive alien species is currently being designed by an expert group under the initiative Implementing European Biodiversity Indicators 2010 (Mace et al., 2005). Given the immediacy of the challenge to meet the 2010 target, broad discussion of the

CBD headline indicator "trends in invasive alien species" and an evaluation of the scientific criteria underlying it are crucial.

Manchester & Bullock (2000) precisely reviewed the status, spread and policy/control legislations pertaining to invasive species. Hughes & Madden (2003) described the use of receiver operating characteristic methodology to evaluate the implications of the different types of regulatory policy for invasive weeds. A risk assessment system was developed to assess the invasion potential of new environmental weeds in central Europe (Weber & Gut 2004).

Biofuel species e.g., *Sorgum halepense, Arundo donax* (giant reed), *Phalaris arundinacea* (reed canary grass), *Panicum virgatum* (hybrid grass) are beneficial in one way, however, act as potential invasive species, therefore, their ecological risk should be assessed before global trade (Raghu et al., 2007; Richardson & Blanchard, 2011).

Mcgeoch et al. (2006) used the CBD 2010 target and headline indicator framework (UNEP, 2005; CBD, 2006) to develop a rationale for the form and characteristics of an indicator of trends in IAS that will meet the 2010 framework goal and targets for this indicator.

Most importantly, economic evaluation studies of biological invasions and of measures against them can be used to support the decision process as a tool of policy advice (Born et al., 2005).

Conclusion

A principal result of the Industrial Revolution and subsequent technological development is the evolution of a planet where the dynamics of major natural systems are increasingly affected by human activity (Allenby, 2008). Invasive species are a serious problem for the world, both ecologically and economically. The impact of invasive species on native species and ecosystems has been immense (Vitousek et al., 1996; Didham et al., 2005). Literatures revealed that 10% of the 260,000 vascular plant species is assumed to be potential invaders (Rapoport, 1991; Sharma et al., 2005). Eighty conifer taxa (79 species and one hybrid; 13% of species) are known to be naturalized, and 36 species (6%) are 'invasive' (Richardson & Rejmánek, 2004).

Majority of research studies on biological invasions is reported from developed countries (McNeely et al., 2009; Khuroo et al., 2011), with scanty information on plant invasions in the biodiversity-rich developing countries, including India (Nunez & Pauchard, 2010; Khuroo et al., 2011), and these gaps have serious implications for global environmental policy making (Khuroo et al., 2011). Towards filling these knowledge gaps, categorization of invasive alien biota is the crucial starting point in developing countries (Khuroo et al., 2011). With reference to developing countries like India having four biodiversity hotspots, developing the consolidated strategy in order to combat the problems of plant invasion is the need of the hour. The conservation, enhancement and sustainable and equitable use of biodiversity should be accorded high priority in all national environment protection programmes (Swaminathan, 2003). In the recently held CABI (Commonwealth Agricultural Bureau International)-MSSRF Workshop on Alien Invasive species at Chennai, India the seven point action plan was developed to manage the invasive alien species (Swaminathan, 2003) with a focus on public awareness; research to fill gaps in our knowledge of the invasive alien species and measure the social, economic and ecological impact of invasive alien species, as well as to evaluate traditional ecological knowledge at the local level; Action at local, state level., national level, regional level and global level should be taken in this regard. Scientific literature on biological invasions in the developing world is currently scarce (Khuroo et al., 2011). India, a fast-globalizing country, faces a high risk of biological invasions. However, research and policy efforts on biological invasions in India are presently inadequate (Khuroo et al., 2011).

In order to address the scale and impacts of this anthropogenic mixing of biotas, as well as an opportunity for basic biological insight, invasion biology has become a rapidly developing discipline with broad ecological and conservation implications (Elton, 1958; Vermeij, 1996; Williamson, 1996; Carlton, 1999; Crooks, 2002). Preparing a comprehensive database of alien floras has been an essential tool for the understanding of plant invasions in different parts of world (Pickard, 1984; Kloot, 1987; Corlett, 1988; Fensham & Cowie, 1998; Rozefelds et al., 1999; Vilá & Muñoz, 1999; Stadler et al., 2000; Pysek et al., 2004; Wu et al., 2004). Biodiversity threats are not evenly distributed, therefore, prioritization is essential to minimize biodiversity loss (Brooks et al., 2006) arising out of the invasion.

Invasion papers were least likely to be cross-linked (6%) with other fields, whereas gap/patch dynamics papers were most likely to be cross-linked (15% (Davis et al., 2005). This tendency toward intellectual isolation may be impeding efforts to achieve more powerful generalizations in ecology by reducing the number of potentially productive exchanges among researchers (Davis et al., 2005).

Furthermore, there is need for greater vigil against alien invasive species, since with growing world trade in food grains and other agricultural commodities, there is an increasing possibility of introducing new pests, weeds and harmful micro-organisms (Swaminathan, 2003). Recent methods based on analyses of genetic markers have provided tools for the retracing of invasion routes and the history of the invading populations from their geographic origin to their final spread in the invaded area (Guillemaud et al., 2011). Guillemaud et al. (2011) suggested that the reconstruction of invasion routes with population genetics-based methods can address fundamental questions in ecology and practical aspects of the management of biological invasions in agricultural settings.

Invasion biologists across the different countries developed different model frameworks for the invasion process in conjunction with different taxa and different environments leading to a plethora of confusing concepts. Therefore, a unified framework for biological invasions that reconciles and integrates the key features of the most commonly used invasion frame- works into a single conceptual model that can be applied to all human-mediated invasions (Blackburn et al., 2011).

References

Achard, F., Eva, H.D., Stibig, H-J., et al. 2002. Determination of Deforestation Rates of the World's Humid Tropical Forests. *Science* 297: 999-1002.

Adair, R.J., Groves, RH. 1998. *Impact of environmental weeds on biodiversity: a review and development of methodology.* Environment Australia, Canberra.

Adams, R.P., Flournoy, LE., Singh, RL., Johnson, H., Mayeux, H. 1998. Invasion of grasslands by *Juniperus ashei*: A new theory based on random amplified polymorphic DNAs (RAPDs). *Biochemical systematic and Ecology* 26: 371-377.

Agarwal, AR., Gahlot, A., Verma, R., Rao, PB. 2002. Effects of weed extracts on seedling growth of some varieties of wheat. *J. Environ. Biol.* 23: 19—23.

Ahmad, R., Liow, P., Spencer, DF., Jasieniuk, M. 2008. Molecular evidence for a single genetic clone of invasive *Arundo donax* in the United States. *Aquatic Botany* 88: 113–120.

Allaie, RR., Reshi, Z., Rashid, I., Wafai., BA. 2006. Effect of Aqueous Leaf Leachate of *Anthemis cotula* – An Alien Invasive Species on Germination Behaviour of Some Field Crops. *J. Agronomy & Crop Science* 192: 186—191.

Allee, WC. 1931. *Animal aggregations: a study in general sociology.* Chicago, IL, USA: University of Chicago Press.

Allen et al. 2004. Host-parasite co-evolutionary conflict between *Arabidopsis* and Downy Mildew. *Science* 306: 1957-1960.

Allenby, B. 2008. The Anthropocene as Media Information Systems and the Creation of the Human Earth. *American Behavioral Scientist* 52 (1): 107-140.

Alpert, P., Bone, E., Holzapfel, C. 2000. Invasiveness, invasibility and the role of environmental stress in the spread of non-native plants. Perspectives in Plant Ecology, *Evolution and Systematics,* Vol. 3/1: 52-66.

Andersen, M.C., Ewald, M., Northcott, J. 2005. Risk analysis and management decisions for weed biological control agents: Ecological theory and modeling results. *Biological Control* 35: 330–337.

Andow, DA., Kareiva, PM., Levin, SA., Okubo, A. 1990. Spread of invading organisms. Landscape *Ecol.* 4: 177–188.

Andrew, N.L., Viejo, R.M. 1998. Ecological limits to the invasion of *Sargassum muticum* in northern Spain. *Aquatic Botany* 60: 251–263.

Angold, PG. 1997. The impact of a road upon adjacent heathland vegetation: effects on plant species composition. *Journal of Applied Ecology* 34: 409–417.

Anning, AK., Yeboah-Gyan, K. 2007. Diversity and distribution of invasive weeds in Ashanti Region, Ghana. *Afr. J. Ecol.,* 45: 355–360.

Anon. 1998. War declared on aliens. *Science* 281: 761.

Antonio, C.D., Meyerson, LA. 2002. Exotic Plant Species as Problems and Solutions in Ecological Restoration: A Synthesis. *Restoration Ecology* 10(4): 703–713.

Antonovics, J., Levin, DA. 1980. The ecological and genetic consequences of density-dependent regulation in plants. *Annual Review of Ecology and Systematics* 11: 411–452.

Arim, M., Abades, S.R., Neill, PE., Lima, M., Marquet, PA. 2006. Spread dynamics of invasive species. *Proc. Nati. Acad Sci* 103 (2), 374-378.

Asner, G.P., Vitousek, P.M. 2005. Remote analysis of biological invasion and biogeochemical change. *Proc. Nati. Acad Sci* 102 (12): 4383–4386.

Augustine, D.J., McNaughton, S.J.. 1998. Ungulate effects on the functional species composition of plant communities: herbivore selectivity and plant tolerance. *J. Wildl. Manage.* 62: 1165–1183.

Baars, J.B. 2003. Geographic range, impact, and parasitism of lepidopteran species associated with the invasive weed *Lantana camara* in South Africa. *Biological Control* 28: 293–301.

Badano, E.I., Pugnaire, FI. 2004. Invasion of *Agave* species (Agavaceae) in south-east Spain: invader demographic parameters and impacts on native species. *Diversity and Distribution* 10: 493–500.

Bais, H.P., Vepachedu, R., Gilroy, S., Callaway, R.M., Vivanco, JM. 2003. Allelopathy and Exotic Plant Invasion: From Molecules and Genes to Species Interactions. *Science* 301: 1377.

Baker, HG. 1955. Self compatibility and establishment of long distance dispersal. *Evolution,* 9: 337–349.

Baker, HG. 1965. *Characteristics and modes of origins of weeds. The genetics of colonizing species* (ed. by H.G. Baker and G.L. Stebbins), pp. 147–172. Academic Press, New York.

Baker, HG. 1967. Support for Baker's law -as a rule. *Evolution* 21: 853–856.

Baker, HG. 1974. The evolution of weeds. *Annual Review of Ecology and Systematics* 7: 1–24.

Bakker, JP. Berendse, F. 1999. Constraints in the restoration of ecological diversity in grassland and heathland communities. *Trends in Ecology and Evolution* 14: 63–68.

Baret et al. 2006. Current distribution and potential extent of the most invasive alien plant species on La Réunion (Indian Ocean, Mascarene islands). *Austral Ecology* 31: 747–758.

Barlow, C. 1997. *Green space, green time.* New York: Springer-Verlag.

Barney, J.N., Tomasso, AD. & Weston LA. 2005. Differences in invasibility of two contrasting habitats and invasiveness of two mugwort *Artemisia vulgaris* populations. *Journal of Applied Ecology* 42: 567–576.

Bastl, M., Kocár, P., Prach, K., Pysek, P. 1997. The effect of successional age and disturbance on the establishment of alien plants in man-made sites: an experimental approach. *Plant Invasions: Studies from North America and Europe* (eds J.H. Brock, M. Wade, P. Pysek & D. Green): pp. 191–201. Backhuys, Leiden.

Bax, N. et al. 2001. The control of biological invasions in the world's oceans. *Conservation Biology* 15 (5): 1234-1246.

Beater, MMT., Garner, R.D., Witkowski, E.T.F. 2008. Impacts of clearing invasive alien plants from 1995 to 2005 on vegetation structure, invasion intensity and ground cover in a temperate to subtropical riparian ecosystem. *South African Journal of Botany* 74: 495–507.

Bellingham, P.J., Duncan, R.P., Lee, W.G., Buxton, RP. 2004. Seedling growth rate and survival do not predict invasiveness in naturalized woody plants in *New Zealand. Oikos* 106: 308-316.

Benning, TL., LaPointe, D., Atkinson, CT., Vitousek, PM. 2002. Interactions of climate change with biological invasions and land use in the Hawaiian Islands: Modeling the fate of endemic birds using a geographic information system. *Proc. Nati. Acad Sci* 99 (22): 14246–14249.

Bergelson, J., Newman, JA. Floresroux, EM. 1993. Rates of weed spread in spatially heterogenous environments. *Ecology* 74: 999-1011.

Bhushan, L., Sharma, P.K. 2005. Long-term effects of lantana residue additions on water retention and transmission properties of a medium-textured soil under rice–wheat cropping in northwest India. *Soil Use and Management* 21: 32–37

Biswas, S.R., Choudhury, J.K., Nishat, A., Rahman, M. 2007. Do invasive plants threaten the Sundarbans mangrove forest of Bangladesh? *Forest Ecology and Management* 245: 1–9.

Blackburn et al. 2011. A proposed unified framework for biological invasions. *Trends in Ecology and Evolution* 26 (7):333-339.

Blondel, J., Aronson, J. 1999. *Biology and Wildlife of the Mediterranean Region.* Oxford

Blossey, B., Nötzold, R. 1995. Evolution of increased competitive ability in invasive nonindigenous plants: a hypothesis. *J. Ecol.* 83: 887–889.

Blumenthal, D., Mitchell, C.E., Pysek, P., Jarosik, V. 2009. Synergy between pathogen release and resource availability in plant invasion. *Proc. Nati. Acad Sci USA* 106: 7899-7904.

Blumenthal, D. 2005. Interrelated Causes of Plant Invasion. *Science* 310: 343-344.

Blumenthal, D. 2006. Interactions between resource availability and enemy release in plant invasion. *Ecology Letters* 9: 887–895.

Blumenthal, D.M., Jordan, NR. & Russelle, MP. 2003. Soil carbon addition controls weeds and facilitates prairie restoration. *Ecol. Appl.* 13: 605–615.

Bohlen, P.J., Groffman, P.M., Fahey, T.J., Fisk, M.C., Sua´rez, E., Pelletier, D.M., Fahey, RT. 2004a. Ecosystem consequences of exotic earthworm invasion of north temperate forests. *Ecosystems* 7: 1–12.

Bohlen, P.J., Scheu, S., Hale, C.M., McLean, MA., Migge, S., Groffman, P.M., Parkinson, D. 2004b. Invasive earthworms as agents of change in north temperate forests. *Front. Ecol. Environ.* 2: 427–435.

Bohlen, PJ. 2006. Biological invasions: Linking the aboveground and belowground consequences. *Applied Soil Ecology* 32: 1–5.

Born, W., Rauschmayer, F., Bra¨uer, I. 2005. Economic evaluation of biological invasions—a survey. *Ecological Economics* 55: 321– 336.

Bradshaw, AD. 1965. Evolutionary significance of phenotypic plasticity in plants. *Advances in Genetics* 13: 115–155.

Briese, D T. 1993. *The Contribution of Plant Biology and Ecology to the Biological Control of Weeds.* Proceedings 10th Australian Weeds Conference and 14th Asian-Pacific Weed Science Society Conference, pp. 10–18, *The Weed Society of Queensland on behalf of The Council of*

Australian Weed Science Societies and the Asian-PacificWeed Science Society, Brisbane.

Briese, DT. 1996. Biological Control of weeds and fire management in protected natural areas: Are they compatible strategies? *Biological Conservation* 77: 135-141.

Bright, C. 1998. *Life Out of Bounds. Bioinvasion in a Borderless World.* W.W. Norton & Company, New York.

Brock, JH., Wade, M., Pysek, P., Green, D. 1997. *Plant Invasions: Studies from North America and Europe.* Backhuys, Leiden.

Brooker, R.W., Callaghan, TV. 1998. The balance between positive and negative plant interactions and its relationship to environmental gradients: a model. *Oikos* 81: 196–207.

Brooker, RW. 2006. Plant–plant interactions and environmental change. New Phytologist 171: 271–284.

Brooks et al. 2006. Global Biodiversity Conservation Priorities. *Science* 313: 58-61.

Brooks, ML. 2003. Effects of increased soil nitrogen on the dominance of alien annual plants in the Mojave Desert. *Journal of Applied Ecology* 40: 344–353.

Brown, JH., Whitham, TG., Ernest, MSK., Gehring, CA. 2001. Dynamics of Ecological Systems: Long-Term Experiments. *Science* 293 (27): 643-650.

Brown, K.A., Gurevitch, J. 2003. Long-term impacts of logging on forest diversity in Madagascar. *Proc. Nati. Acad Sci* 101 (16): 6045–6049.

Buckley et al. 2006. Management of plant invasions mediated by frugivore interactions. *Journal of Applied Ecology* 43: 848–857.

Buckley, LB., Roughgarden, J. 2004. Effects of changes in climate and land use. *Nature* 430: U2.

Buckley, YM., Anderson, S., Catterall, CP. et al. (2006) Management of plant invasions mediated by frugivore interactions. *Journal of Applied Ecology* 43 (5): 848-857.

Bunker, DE. et al. 2005. Species Loss and Aboveground Carbon Storage in a Tropical Forest. *Science* 310: 1029-1031.

Burgess, TL. Bowers, JE., Turner, RM. 1991. *Exotic plants of the desert laboratory,* Tucson, Arizona. Madrono. 38: 96-114.

Bush, MB. Silman MR., Urrego, DH. 2004. 48,000 Years of Climate and Forest Change in a Biodiversity Hot Spot. *Science* 303: 827-829.

Byers, J.E. et al. 2001. Directing research to reduce the impacts of nonindegenous species. *Conservation Biology* 16 (3): 630-640.

Cabin, RJ., Weller, SG., Lorence, DH., Cordell, S., Hadway, LJ., Montgomery, R., Goo D, Urakami, A. 2002. Effects of light, alien grass and native species additions on Hawaiian dry forest restoration. *Ecological Applications* 12: 1595–1610.

Cacho, OJ. et al. 2007. Bioeconomic modeling for control of weeds in natural environments. *Ecological Economics* 65 (3): 559-568.

Cadenasso, ML., Pickett, STA. 2001. Effects of edge structure on the flux of species into forest interiors. *Conserv. Biol.* 15: 91–97.

Callaway, et al. 2008. Novel weapons: Invasive plant suppresses fungal mutualists in America but not in its native. *Europe. Ecology,* 89(4): 1043–1055.

Callaway, R.G., Aschehoug, ET. 2000. Invasive plants versus their new and old neighbors: a mechanism for exotic invasion. *Science* 290: 521–523.

Callaway, R.G., Ridenour, W.M., Laboski, T., Weir, T., Vivanco, JM. 2005. Natural selection for resistance to the allelopathic effects of invasive plants. *Journal of Ecology* 93: 576–583.

Callaway, R.M., Ridenour, W.M. 2004. Novel weapons: invasive success and the evolution of increased competitive ability. Front Ecol Environ 2(8): 436–443.

Callaway, R.M., Thelen, GC., Rodriguez, A., Holben, WE. 2004. Soil biota and exotic plant invasion. *Nature* 427: 731-733.

Callaway, RM., & Maron, JL. 2006. What have exotic plant invasions taught us over the past 20 years? *Trends Ecol. Evol.* 21(7), 369-374.

Callaway, RM. 1995. Positive interactions among plants. *Botanical Review* 61: 306–349.

Callaway, RM. 2002. The detection of neighbors by plants. *Trends Ecol. Evol.* 17, 104.

Carlton, J T. 1999. A journal of biological invasions. *Biol. Invasions* 1: 1.

Carlton, JT. 1989. Man's Role in Changing the Face of the Ocean Biological Invasions and Implications for Conservation of Near-Shore Environments. *Conserv. Biol.* 3: 265.

Carlton, JT. 1996. Biological invasions and cryptogenic species. *Ecology* 77: 1653–1655.

Case, TJ. 1990. Invasion resistance arises in strongly interacting species-rich model competition communities. *Proc. Natl Acad. Sci.* 87: 9610–9614.

Case, TJ. 1987. Travelers and Their Fate (Book Reviews*). Science* 236, 1000-1002.

Chapin, FS., Zavaleta, ES., Eviner, VT., Naylor, RL., Vitousek, PM. 2000. Consequences of changing biodiversity. *Nature* 405, 234-242.

Chapin III, FS., Walker, BH., Hobbs, RH., Hooper, DU, Lawton, JH., Sala, OE, Tilman, D. 1997. Biotic Control over the Functioning of Ecosystems. *Science* 277: 500-504.

Chapuis, JL., Bousse's, P, Barnaud G. 1994. Alien mammals, impact and management in the French subantarctic islands. *Biol. Conserv.* 67: 97–104.

Chin, G. 2005. The Dynamics of Invasions (Editor's Choice). *Science* 310: 747.

Chown, SL., Rodrigues, AS., Gremmen, NJM., Gaston KJ. 2001. World Heritage status and the conservation of Southern Ocean islands. *Conservation Biology* 15: 550–557.

Claassen, VP., Marler, M. 1998 Annual and perennial grass growth on nitrogen-depleted decomposed granite. *Restoration Ecology* 6: 175–180.

Clay, K., Holah, J., Rudgers, JA.2005. Herbivores cause a rapid increase in hereditary symbiosis and alter plant community composition. *Proc. Nati. Acad Sci* 102 (35): 12465-12470.

Coblentz, BE. 1978. The effects of feral goats (*Capra hircus*) on island ecosystems. – *Biol. Conserv.* 13: 279–286.

Cohen, AN., Carlton, JT. 1998. Accelerating Invasion Rate in a Highly Invaded Estuary. *Science* 279: 555.

Colautti, RI., MacIsaac, HJ. 2004. A neutral terminology to define 'invasive' species. *Diversity and Distributions* 10: 135–141.

Colnar, AM., Wayne, G. 2007. Conceptual Model Development for Invasive Species and a Regional Risk Assessment Case Study: The European Green Crab, *Carcinus maenas*, at Cherry Point, Washington, USA. *Human and Ecological Risk Assessment* 13: 120-155.

Constanza, R., d'Arge R, de Groot R, Farber S, Grasso M, Hannon B, Limburg K, Naeem S, O'Neill RV, Paruelo J, Raskin RG, Sutton P, van den Belt, M. 1997. The value of the world's ecosystem services and natural capital. *Nature* 387, 253–260.

COP. 2002. Review and consideration of options for the implementation of Article 8(h) on alien species that threatens ecosystems, habitats or species. *Conference of the parties to the Convention on Biological Diversity (COP).* UNEP/CBD/COP/6/18/Add.1/ Rev.1. 26-3-2002.

Corlett, RT. 1988. The naturalized flora of Singapore. *Journal of Biogeography* 15: 657–663.

Costanza, R. et al. 1997. The value of the world's ecosystem services and natural capital. *Nature* 387: 253–260.

Costello, DA., Lunt,,ID. Williams JE. 2000. Effects of invasion by the indigenous shrub *Acacia sophorae* on plant composition of coastal grasslands in south-eastern Australia. *Biological Conservation* 96: 113-121.

Crawley, MJ., Brown SL, Heard MS, Edwards GR. 1999. Invasion-resistance in experimental grassland communities: species richness or species identity? *Ecology Letters* 2: 140–148.

Crawley, MJ. 1987. *What makes a community invasible?* In: A.J. Gray, M.J. Crawley and P.J. Edwards (Editors), *Colonization, Succession and Stability.* Blackwell, Oxford, pp. 429-453.

Crawley, MJ. 2005. Learning from the aliens. *Science* 310: 623-624.

Crawley, M. J. 1987. *Colonization, Succession and Stability* (eds Crawley,M. J., Edwards, P. J. & Gray, A. J.) Blackwell, Oxford:pp 429–453.

Cronk, QC.B., Fuller JL. 1995. *Plant invaders – the threat to natural ecosystems.* Chapman & Hall, London.

Crooks, JA. 2002. Characterizing ecosystem-level consequences of biological invasions: the role of ecosystem engineers. *Oikos* 97: 153–166.

Crossman, ND., Bryan, BA., Cooke, DA. 2011. An invasive plant and climate change threat index for weed risk management: Integrating habitat distribution pattern and dispersal process. *Ecological Indicators* 11: 1 8 3 – 1 9 8.

D'Antonio, CM., Vitousek, PM. 1992 Biological invasions by exotic grasses, the grass/fire cycle, and global change. *Annu Rev Ecol System* 23:63–87.

D'Antonio, CM., Hobbie, SE. 2005. Plant species effects on ecosystem processes. In: Sax DF, Stachowicz JJ, Gaines SD, eds. *Species invasions: insights from ecology, evolution and biogeography.* Sunderland, MA, USA: Sinauer Associates, 65–84.

D'Antonio, C.M., Kark, S. 2002. Impacts and extent of biotic invasions in terrestrial ecosystems (Meeting report). *Trends Ecol. Evol.* 17 (5): 202-204.

D'Antonio, C.M., Mahall, BE. 1991. Root profiles and competition between the invasive, exotic perennial, Carpobrotus edulis, and two native shrub species in California coastal scrub. American Journal of Botany 78: 885–894.

D'Antonio, C.M., Tunison, J.T., Rhondak, KL. 2000. Variation of the impact of exotic grasses on native plant composition in relation to fire across an elevation gradient in Hawaii. *Austral Ecology* 25: 507–522.

D'Antonio, CM., Vitousek, PM. 1992. Biological invasions by exotic grasses, the grass/fire cycle, and global change. *Annual Review of Ecology and Systematics* 23: 63–87.

Daehler, CC. 1998. The taxonomic distribution of invasive angiosperm plants: Ecological insights and comparison to agricultural weeds. *Biological Conservation* 84: 167-I80.

Daehler, CC. 2003. Performance comparisons of co-occurring native and alien invasive plants: implications for conservation and restoration. *Annu. Rev. Ecol. Syst.* 34: 183–211.

Daehler, CC. 2005. Upper-montane plant invasions in the Hawaiian Islands: Patterns and opportunities. *Perspectives in Plant Ecology, Evolution and Systematics* 7: 203–216.

Daily, GC. 1997. *Nature's Services: Societal Dependence on Natural Ecosystems.* Island Press, Washington, DC.

Damschen et al. 2006. Corridors Increase Plant Species Richness at Large Scales. *Science* 313: 1284-1286.

D'Antonio, CM. 1993. Mechanisms controlling invasion of coastal plant communities by the alien succulent *Carpobrotus edulis. Ecology* 74: 83-95.

Dark, SJ. 2004. The biogeography of invasive alien plants in California: an application of GIS and spatial regression analysis. *Diversity Distrib.* 10: 1-9.

Darwin, C. 1859. *On the Origin of Species by Means of Natural Selection, or the Preservation of Favoured Races in the Struggle for Life*, John Murray, London, pp.116

Date, EM., Recher, HF, Ford HA. Stewart, DA. 1996. The conservation and ecology of rainforest pigeons in north-eastern New South Wales. *Pacific Conservation Biology* 2: 299–308.

Davis, AJ., Jenkinson, LS., Lawton, JH., Shorrocks, B., Wood, S. 1998. Making mistakes when predicting shifts in species range in response to global warming. *Nature* 391: 783–786.

Davis, AM. 2009. *Invasion Biology.* Oxford University Press. pp. 244.

Davis et al. 2005. Vegetation change: a reunifying concept in plant ecology. *Perspectives in Plant Ecology, Evolution and Systematics* 7: 69–76.

Davis, H.G., Taylor, C.M., Civille, JC., Strong, DR. 2004. An Allee effect at the front of a plant invasion: *Spartina* in a Pacific estuary. *Journal of Ecology* 92: 321–327.

Davis, MA., Grime, JP.. & Thompson, K. 2000. Fluctuating resources in plant communities: a general theory of invisibility. *Journal of Ecology* 88: 528-534.

Davis, MA., Wrage, KJ., Reich, PB., Tjoelker, MG., Schaeffer, T., Muermann, C. 1999. Survival, growth, and photosynthesis of tree seedlings competing with herbaceous vegetation along a water-light-nitrogen gradient. *Plant Ecology* 145: 341-350.

Davis, MA., Wrage, KJ. Reich, PB. 1998. Competition between tree seedlings and herbaceous vegetation: support for a theory of resource supply and demand. *Journal of Ecology* 86: 652-661.

de Groot M, Kleijna D, Jogan N. 2007. Species groups occupying different trophic levels respond differently to the invasion of semi-natural vegetation by *Solidago Canadensis*. *Biological Conservation* 136: 612-617.

de Villalobos AE, et al. 2011. *Pinus halepensis* invasion in mountain pampean grassland: Effects of feral horses grazing on seedling establishment. *Environ. Res.* 111(7): 953-959..

de Waal LC, Child LE, Wade PM, Brock JH. 1994. *Ecology and Management of Invasive Riverside Plants*. Wiley, Chichester.

Dean, WRJ. 1998. Space invaders: modelling the distribution, impacts and control of alien organisms. *Trends Ecol. Evol.* 13: 256-258.

DeFarrari, CM., Naiman, RJ., 1994. A multi-scale assessment of the occurrence of exotic plants on the Olympic Peninsula, Washington. *J. Veg. Sci.* 5: 247-258.

Dehnen-Schmutz, K. 2004. Alien species reflecting history: medieval castles in Germany. *Diversity and Distributions* 10: 147–151.

Di Tomaso, JM. 1998. Impact, biology, and ecology of saltcedar (*Tamarix* spp.) in the south-western United States. *Weed Technology* 12: 326–336.

Didham, RK., Tylianakis, JM., Hutchison MA, Ewers RM, Gemmell, NJ. 2005. Are invasive species the drivers of ecological change? *Trends Ecol. Evol.* 20(9): 470-474.

Dingwall, P R. 1995. *Progress in conservation of the subantarctic islands.* IUCN, Gland.

DiTomaso, JM. 2000. Invasive weeds in rangelands. Species, impacts, and management. *Weed Science* 48: 255–265.

Donlan, C.J. et al. 2003. Research for requiems: the need for more collaborative action in eradication of invasive species. *Conservation Biology* 17(6): 1850-1851.

Dorken, ME. Barrett, SCH. 2004. Phenotypic plasticity of vegetative and reproductive traits in monoecious and dioecious populations of *Sagittaria latifolia (Alismataceae)*: a clonal aquatic plant. *Journal of Ecology* 92: 32–44.

Drake, JA, Mooney HA, Di Castri F, Groves, RH, Kruger, FJ, Rejma´nek, M, Williams M. (Eds.), 1989. *Biological Invasions: A Global Perspective.* John Wiley and Sons, Chichester, UK.

Driesche, JV., Driesche, RV. 2000. Nature out of Place. *Biological Invasions in the Global Age.* Island Press, pp. 352.

Duggin, JA. Gentle CB. 1998. Experimental evidence on the importance of disturbance intensity for invasion of *Lantana camara* L. in dry rainforest-open forest ecotones in north-eastern NSW, Australia. *Forest Ecology and Management* 109: 279-292.

Dukes, JS., Mooney, HA. 1999. Does global change increase the success of biological invaders? *Trends Ecol. Evol.* 14:135-139.

Dukes, JS., Mooney, H.A. 2004. Disruption of ecosystem processes in western North America by invasive species. *Revista Chilena de Historia Natural* 77: 411–437.

Dukes, JS. 2001 Biodiversity and invasibility in grassland microcosms. *Oecologia* 126: 563–568.

Dukes, JS. 2002. Comparison of the effect of elevated CO_2 on an invasive species (*Centaurea solstitialis*) in monoculture and community settings. *Plant Ecology* 160: 225–234.

Dynesius, M., Jansson, R. 2000. Evolutionary consequences of changes in species' geographical distributions driven by Milankovitch climate oscillations. *Proc. Nati. Acad Sci* 97 (16): 9115–9120.

Eckert, CG., Manicacci, D., Barrett, SCH. 1996. Genetic drift and founder effect in native versus introduced populations of an invading plant, *Lythrum salicaria* (Lythraceae). *Evolution,* 50: 1512–1519.

Edwards, C.A. 2000. Soil invertebrate controls and microbial interactions in nutrient and organic matter dynamics in natural and agroecosystems. In: Coleman, D. C., Hendrix, P. F. (eds) *Invertebrates as webmasters in ecosystems.* CABI Publishing, Wallingford, pp. 141–159.

Ehrenfeld, JG. 2010. Ecosystem Consequences of Biological Invasions. *Annu. Rev. Ecol. Evol. Syst.* 41: 59–80.

Elam, DR., Ridley, CE., Goodell, K., Ellstrand, NC. 2007. Population size and relatedness affect fitness of a self-incompatible invasive plant. *Proc. Nati. Acad Sci* 104 (2): 549–552.

El-Ghareeb, R. 1991. Vegetation and soil changes induced by *Mesembryanthemum crystallinum* L. in a Mediterranean desert ecosystem. *J. Arid Environ.* 20: 321–330.

Ellison, CA., Evans, HC., Djeddour, DH., Thomas, S.E. 2008. Biology and host range of the rust fungus *Puccinia spegazzinii*: a new classical biological control agent for the invasive, alien weed *Mikania micrantha* in Asia. *Biological Control* 45(1): 133-145.

Ellstrand. N.C., Schierenbeck, K.A. 2000. Hybridization as a stimulus for the evolution of invasiveness in plants? *Proc. Nati. Acad Sci* 97 (13): 7043–7050.

Elton, C. S. 1958. *The ecology of invasions by animals and plants.* – John Wiley and Sons.

Elton, C. S. 1927 *Animal Ecology* Sidgewick and Jackson, London,.

Enserink, M. 1999. Biological Invaders sweep in. *Science* 285: 1834-36.

ESA (Ecological Society of America) 2004. Are Invasive Species Born Bad? *Meeting Ecological Society of America Science,* 20 August, 305.

Feng, Z., Qiu, Z., Liu, R., DeAngelis, DL. 2011. Dynamics of a plant–herbivore–predator system with plant-toxicity. *Mathematical Biosciences* 229: 190–204.

Fensham, RJ., & Cowie, I.D. 1998. Alien plant invasions on the Tiwi islands: extent, implications and priorities for control. *Biological Conservation* 83: 55–68.

Field, B., Jordán, F., Osbourn, A. 2006. First encounters – deployment of defence-related natural products by plants. *New Phytologist* 172: 193–207.

Filchak, K.E. et al. 2000. Natural selection and sympatric divergence in the apple aggot *Rhagoletis pomonella*. *Nature* 407: 739–742.

Fine, PVA. 2002. The invasability of tropical forests by exotic plants. *J. Trop. Ecol.* 18: 687–705.

Fitter, A. 2003. Making Allelopathy Respectable. *Science* 301: 1337-1338.

Fleishman, E., Mac Nally, R., Murphy, D.D. 2005. Relationships among non-native plants, diversity of plants and butterflies, and adequacy of spatial sampling. *Biological Journal of the Linnean Society* 85: 157–166.

Florentine, S.K., Westbrooke, M.E., Gosney, K., Ambrose, G., O'Keefe, M. 2006. The arid land invasive weed *Nicotiana glauca* R. Graham (Solanaceae): Population and soil seed bank dynamics, seed germination patterns and seedling response to flood and drought. *Journal of Arid Environments* 66: 218–230.

Forman, R.T.T., Alexander, L.E., 1998. Roads and their major ecological effects. *Annual Review of Ecology and Systematics* 29: 207–231.

Forman, R.T.T., Deblinger, R.D., 2000. The ecological road-effect zone of a Massachusetts (USA) suburban highway. *Conservation Biology* 14: 36–46.

Foster, P. 2001. The potential negative impacts of global climate change on tropical montane cloud forests. *Earth-Science Reviews* 55: 73–106.

Fox, M.D. & Fox, B.D. 1986. The susceptibility of natural communities to invasion. In: R.H. Groves and J.J. Burdon (Editors), *Ecology of Biological Invasions: An Australian Perspective. Australian Academy of Science,* Canberra, pp. 97-105.

Foxcroft, L.C., Rouget, M., Richardson, D.M., Mac Fadyen, S. 2004. Reconstructing 50 years of *Opuntia stricta* invasion in the Kruger National Park, South Africa: environmental determinants and propagule pressure. *Diversity and Distributions* 10: 427–437.

Foxcroft, L.C., Hoffmann, J.H., Viljoen, J.J., Kotze, J.J. 2007. Environmental factors influencing the distribution of *Opuntia stricta*, an invasive alien plant in the Kruger National Park, South Africa. *South African Journal of Botany* 73: 109–112.

Foxcroft, L.C., Pickett, S.T.A, Cadenasso, M.L. 2011. Expanding the conceptual frameworks of plant invasion ecology. Perspectives in Plant Ecology, *Evolution and Systematics* 13: 89–100.

Frenot, Y. et al. 2005. Biological invasions in the Antarctic: extent, impacts and implications. Biol. Rev. 80: 45–72.

Funk, J.L., Vitousek, P.M. 2007. Resource-use efficiency and plant invasion in low-resource systems. *Nature* 446: 1079-1081.

Gallagher, R., Carpenter, B. 1997. Human-dominated ecosystems: Introduction. *Science* 277: 485.

Garrity, D.P. et al. 1997. The *Imperata* grasslands of tropical Asia: area, distribution, and typology. *Agrofor. Syst.* 36: 1–29.

Gaskin, J.F., Schaal, B.A. 2002. Hybrid *Tamarix* widespread in U.S. invasion and undetected in native Asian range. *Proc. Nati. Acad Sci* 99 (17): 11256–11259.

Gelbard, J.L., Belnap, J. 2003. Roads as Conduits for Exotic Plant Invasions in a Semiarid Landscape. *Conservation Biology* 17(2): 420–432.

Gentle, C.B., Duggin, JA. 1997. *Lantana camara* L. invasions in dry rainforest - open forest ecotones: The role of disturbances associated with fire and cattle grazing. *Australian Journal of Ecology* 22: 298-306.

Gewin, V. 2005. Industry lured by the gains of going green. *Nature* 436: 173.

Ghazoul, J. 2002. Flowers at the front line of invasion? *Ecol. Entomol.* 27, 638-640.

Ghermandi L, de Torres Curth, MI, Franzese J, Gonzalez S. 2010. Non-linear ecological processes, fires, environmental heterogeneity and shrub invasion in northwestern Patagonia. *Ecological Modelling* 221: 113–121.

GISP, 2000. *Global strategy on invasive alien species. Global Invasive Species Programme workshop*, September 2000, Cape Town, South Africa, 2001.

GISP, 2003. The IAS problem. The Global Invasive Species Programme, 2003 http://www.gisp.org/about/IAS.asp.

Glowka, L., Burhenne-Guilmin, F., Synge, H. 1994. *A guide to the convention on biological diversity.* - IUCN. Gland.

Godefroid, S., Koedam, N. 2004. The impact of forest paths upon adjacent vegetation: effects of the path surfacing material on the species composition and soil compaction. *Biological Conservation* 119: 405–419.

Godfree, R.C., Thrall, P.H., Young, AG. 2006. Enemy release after introduction of disease-resistant genotypes into plant–pathogen systems. *Proc. Nati. Acad Sci* 104 (8): 2756-2760.

Goel, U., Saxena, D.B., Kumar, B., Birendra, K. 1989. Comparative study of allelopathy as exhibited by *Prosopis juliflora* and *Prosopis cineraria. J. Chem. Ecol.* 15: 591—600.

Goodall, J.M., Erasmus, D.J. 1996. Review of the status and integrated control of the invasive alien weed, *Chromolaena odorata,* in South Africa. Agriculture, *Ecosystem and Environment* 56: 15I - I64.

Goodall, J.M., Naude, DC. 1998. An ecosystem approach for planning sustainable management of environmental weeds in South Africa. Agriculture, *Ecosystems and Environment* 68: 109–123.

Gooden, B., French, K., Turner, PJ. 2009. Invasion and management of a woody plant, *Lantana camara* L., alters vegetation diversity within wet sclerophyll forest in southeastern Australia. *Forest Ecology and Management* 257: 960–967.

Goodman, D. 1975.The theory of diversity-stability relationships in ecology. The Quarterly Review of Biology. 50(3):237-266.

Gopal, B., Sharma, K.P. 1981: *Water-Hyacinth: The most Troublesome Weed in the World.* Hindasia Publishers, Delhi.

Goslee, S.C., Peters, D.P.C., Beck, K.G. 2001. Modeling invasive weeds in grasslands: the role of allelopathy in *Acroptilon repens* invasion. *Ecological Modelling* 139: 31–45.

Gosper, C.R., Vivian Smith, G. 2006. Selecting replacements for invasive plants to support frugivores in highly modified sites: A case study focusing on *Lantana camara. Ecological Management and Restoration* 7(3): 197-203.

Gosper, C.R. 2004 *Consequences of Weed Invasion and Control on Plant-Bird Interactions and Bird Communities.* PhD Thesis, Department of Biological Sciences, University of Wollongong, Wollongong, Australia. Available from URL: http://www.library.uow.edu.au/theses/.

Gosper, C.R., Vivian-Smith, G. 2006. Selecting replacements for invasive plants to support frugivores in highly modified sites: A case study focusing on *Lantana camara.* Ecological management & Restoration 7 (3): 197-203.

Goulson, D., Derwent, L.C. 2004. Synergistic interactions between an exotic honeybee and an exotic weed: pollination of *Lantana camara* in Australia. Weed Research 44: 195–202.

Graves, S.D., Shapiro, A.M. 2003 Exotics as host plants of the Californian butterfly fauna. *Biological Conservation* 110: 413–433.

Gravuer, K., Sullivan, J.J., Williams, P.A., Duncan, R.P. 2008. Strong human association with plant invasion success for *Trifolium* introductions to New Zealand. *Proc. Nati. Acad Sci* 105 (17): 6344–6349.

Greenslade, P. 2002. Assessing the risk of exotic Collembola invading subantarctic islands : prioritizing quarantine management. *Pedobiologia* 46: 338–344.

Greenwood, H., O'Dowd, DJ., Lake, P.S. 2004. Willow (*Salix* × *rubens*) invasion of the riparian zone in south-eastern Australia: reduced abundance and altered composition of terrestrial arthropods. *Diversity Distrib.* 10: 485–492.

Grice, AC. 2004. Weeds and the monitoring of biodiversity in Australian rangelands. *Austral Ecology* 29: 51–58.

Grime, JP. 2001. *Plant Strategies, Vegetation Processes and Ecosystem Properties* (Wiley, Chichester, UK).

Grinnell, J. 1917. *The niche-relationships of the California thrasher.* Auk 34: 427–433.

Grinnell, J. 1919. The English sparrow has arrived in Death Valley: an experiment in nature. *Am. Nat.* 53: 468.

Grosholz, ED. 2005. Recent biological invasion may hasten invasional meltdown by accelerating historical introductions. *Proc. Nati. Acad Sci* 102 (4): 1088-1091.

Grosholz, E. 2002. Ecological and evolutionary consequences of coastal invasions. *Trends Ecol. Evol.* 17 (1): 22-27.

Groves, R.H., & Willis, AJ. 1999. Environmental weeds and loss of native plant biodiversity: some Australian examples. *Aust. J. Envir. Manag.* 6: 164–71.

Groves, R.H., Willis, AJ. 1999. Environmental weeds and loss of native plant biodiversity: some Australian examples. *Australian Journal of Environmental Management* 6:164–171.

Gu, W., Sang, W., Liang, H., Axmacher, J.C. 2007. Effects of Crofton weed *Ageratina adenophora* on assemblages of Carabidae (Coleoptera) in the Yunnan Province, South China. *Agriculture, Ecosystems and Environment* 124(3-4):173-178.

Guillemaud, T., Ciosi, M., Lombaert, E., Estoup, A. 2011. Biological invasions in agricultural settings: Insights from evolutionary biology and population genetics. C. R. *Biologies* 334: 237–246.

Gurevitch, J., Padilla, D.K. 2004. Are invasive species a major cause of extinctions? *Trends Ecol. Evol.* 19: 470–474.

Hamada, Y., Stow, D.A., Coulter, L.L., Jafolla, J.C., Hendricks, L.W. 2007. Detecting Tamarisk species (*Tamarix* spp.) in riparian habitats of Southern California using high spatial resolution hyperspectral imagery. *Remote Sensing of Environment* 109: 237–248.

Hänfling, B., Kollmann, J. 2002. An evolutionary perspective of biological invasions. *Trends in Ecol. Evol.* 17(12): 545-546.

Hansen, A.J., Neilson, R.R., Dale, V.H. et al. (2001) Global change in forests: responses of species, communities, and biomes. *Bioscience* 51: 765–779.

Hansen, M.J., Clevenger, AP. 2005. The influence of disturbance and habitat on the presence of non-native plant species along transport corridors. *Biological Conservation* 125: 249–259.

Harrington, GN. 1991. Effects of soil moisture on shrub seedling survival in a semi-arid grassland. *Ecology* 72: 1138-1149.

Harrison, S. 1999. Native and alien species diversity at the local and regional scales in a grazed California grassland. *Oecologia* 121: 99-106.

Hättenschwiler, S., Körner, C. 2003. Does elevated CO_2 facilitate naturalization of the non-indigenous *Prunus laurocerasus* in Swiss temperate forests? *Journal of Ecology* 17: 778–785.

HäXiger, P., Schwarzländer, M., Blossey, B. 2006. Impact of *Archanara geminipuncta* (Lepidoptera: Noctuidae) on aboveground biomass production of *Phragmites australis*. *Biological Control* 38: 413–421.

Hay, ME., Hollebone, A.L. 2007. Propagule pressure of an invasive crab overwhelms native biotic resistance. *Mar Ecol Prog Ser* 342: 191-96.

Herron, P.M., Martine, C.T., Latimer, A.M. Leicht-Young, SA. 2007. Invasive plants and their ecological strategies: prediction and explanation of woody plant invasion in New England. *Diversity Distrib.* 13: 633–644.

Hierro, JL., Callaway, RM. 2003. Allelopathy and exotic plant invasion. *Plant Soil* 256: 29–39.

Higgins, S.I., Richardson, D.M., Cowling, R.M., Trinder-Smith, TH. 1999. Predicting the landscape — scale distribution of alien plants and their threat to plant diversity. *Conservation Biology,* 13: 303–313.

Higgins, S.I., Richardson, D.M. 1996. A review of models of alien plant spread. *Ecological Modelling* 87: 249–265.

Hill, S.J., Tung, P.J., Leishman, R. 2005. Relationships between anthropogenic disturbance, soil properties and plant invasion in endangered Cumberland Plain Woodland, Australia. *Austral Ecology* 30: 775–788.

Hobbs, NT. 1996. Modification of ecosystems by ungulates. *J.Wildl. Manage.* 60: 695–713.

Hobbs, R.J., Humphries, S.E. 1995. An integrated approach to the ecology and management of plant invasions. *Conservation Biology* 9: 761–770.

Hobbs, RJ., Gulman, S.L., Hobbs, V.J., Mooney, HA. 1988. Effects of fertiliser addition and subsequent gopher disturbance on a serpentine annual grassland community. *Oecologia* 75: 291-295.

Hobbs, RJ., Huenneke, LF. 1992. Disturbance, diversity, and invasion: Implications for conservation. *Conserv. Biol.* 6: 324-337.

Hobbs, R.J., Mooney, H.A. 1985. Community and population dynamics of serpentine grassland annuals in relation to gopher disturbance. *Oecologia* 67: 342-351.

Hobbs, R.J., Mooney, H.A. 1991. Effects of rainfall variability and gopher disturbance on serpentine annual grassland dynamics. *Ecology* 72: 59-68.

Hobbs, R.J. 1989. The nature and effects of disturbance relative to invasions. In: J.A. Drake, H.A. Mooney, F. di Castri, R.H. Groves, F.J. Kruger, M. Rejmanek and M. Williamson (Editors), *Biological Invasions: A Global Perspective.* John Wiley and Sons, Chichester, pp. 389-405.

Hobbs, R.J., Atkins, L. 1988. Effect of disturbance and nutrient addition on native and introduced annuals in plant communities in the Western Australia wheatbelt. *Australian Journal of Ecology* 13: 171-179.

Hoffmann, J.H., Impson, F.A.C., Moran, V.C., Donnelly, D. 2002. Biological control of invasive golden wattle trees (*Acacia pycnantha*) by a gall wasp, *Trichilogaster* sp. (Hymenoptera: Pteromalidae), in South Africa. *Biological Control* 25: 64–73.

Hofstra, D.E., Clayton, J., Green, J.D., Adam, K.D. 2000. RAPD profiling and isozyme analysis of New Zealand *Hydrilla verticillata*. *Aquatic Botany* 66: 153–166.

Holdgate, M.W. 1986. Summary and conclusions: characteristics and consequences of biological invasions. *Philos. Trans. R. Soc. Lond. Ser. B Biol. Sci.* 314: 733–742.

Hollebone, A.L., Wiesemeier, T. 2007. An invasive crab alters interaction webs in a marine community. *Biol. Invasions* 10: 347-58.

Holmes, P.M., Richardson, D.M., Esler, K.J., Witkowski, E.T.F., Fourie, S. 2005. A decision-making framework for restoring riparian zones degraded by invasive alien plants in South Africa. *South African Journal of Science* 101: 553–564.

Hood, W.G., & Naiman, R.J. (2000) Vulnerability of riparian zones to invasion by exotic vascular plants. *Plant Ecology* 148: 105–114.

Houlahan, J.E., Findlay, C.S. 2002. Effect of invasive plant species on temperate wetland plant diversity. *Conservation Biology* 1132-1138.

Hoya, A., Shibaike, H., Morita, T., Ito, M. 2004. Germination and seedling survivorship characteristics of hybrids between native and alien species of dandelion (*Taraxacum*). *Plant Species Biology* 19: 81–90.

Huang, Hua-mei, Zhang, Li-quan, Guan, Yu-juan, Wang, Dong-hui. 2008. A cellular automata model for population expansion of *Spartina alterniflora* at Jiuduansha Shoals, Shanghai, China. *Estuarine, Coastal and Shelf Science* 77: 47-55.

Huenneke, L.F., Hamburg, S.P., Koide, R., Mooney, H.A., Vitousek, PM. 1990. Effects of soil resources on plant invasion and community structure in Californian serpentine grassland. *Ecology* 71: 478-471.

Hughes, F., Vitousek, P.M., Tunison, T. 1991. Alien grass invasion and fire in the seasonal submontane zone of Hawai'i. *Ecology* 72: 743–746.

Hughesa, G., Madden, L.V. 2003. Evaluating predictive models with application in regulatory policy for invasive weeds. *Agricultural Systems* 76: 755–774

Hulme, P.E., Bremner, E.T. 2006. Assessing the impact of *Impatiens glandulifera* on riparian habitats: partitioning diversity components following species removal. *Journal of Applied Ecology* 43: 43–50.

Hulme, P.E., Pysek, P., Nentwig, W., t Vilà, M. 2009. Will Threat of Biological Invasions Unite the European Union? *Science* 324: 40-41.

Hulme, PE. 2006. Beyond control: wider implications for the management of biological invasions. *Journal of Applied Ecology* 43: 835–847.

Hunt, ER., McMurtrey, JE., Parker, Williams, A.E., Corp, LA. 2004. Spectral caracteristics of leafy spruge leaves and flower bracts *Weed Sci.* 52: 492–497.

Huston, M.A. 2004. Management strategies for plant invasions: manipulating productivity, disturbance, and competition. *Diversity Distrib.* 10: 167–178.

Idso, SB. 1992. Shrubland expansion in the American southwest. *Climate Change* 22: 85-86.

Inderjit et al. 2005. Challenges, achievements and opportunities in allelopathy research. *J. Plant Interac.* 1: 69–81.

Inderjit, Callaway, R.M., Vivanco, J.M. 2006. Can plant biochemistry contribute to understanding of invasion ecology? *Trends in Plant Science* 11(12): 574-580.

Inderjit, Duke, SO. 2003. Ecophysiological aspects of allelopathy. *Planta* 217: 529–539.

Inderjit, Keating, K.I. 1999. Allelopathy: principles, procedures, processes, and promises for biological control. *Adv. Agron.* 67: 141–231.

Ingham, R.E., Trofymow, J.A., Ingham, E.R., Coleman, DC. 1985. Interactions of bacteria, fungi, and their nematode grazers: effects on nutrient cycling and plant growth. *Ecological Monographs* 55:119–140.

Ives, A.R., Carpenter, S.R. 2007. Stability and Diversity of Ecosystems. *Science* 317: 58-62.

Jablonski, D. 1991. Extinctions: a paleontological perspective. *Science* 253: 754-757.

Jager, H., Tye, A., Kowarik, I. 2007. Tree invasion in naturally treeless environments: Impacts of quinine (*Cinchona pubescens*) trees on native vegetation in Gala'pagos. *Biological Conservation* 140: 297-307.

Jakobs, G., Weber, E., Edwards, P.J. 2004. Introduced plants of the invasive *Solidago gigantean* (Asteraceae) are larger and grow denser than conspecifics in the native range. *Diversity and Distributions* 10: 11–19.

Järemo, J., Bengtsson, G. 2011. On the importance of life history and age structure in biological invasions. *Ecological Modelling* 222: 485–492.

Jay, M., Morad, M., Bell, A. 2003. Biosecurity, a policy dilemma for New Zealand. *Land Use Policy* 20: 121– 129.

Jetz, W. Rahbek, C. 2002. Geographic range size and determinants of avian species richness. *Science* 297: 1548–1551.

Johnson, H.B., Polley, H.W., Mayeux, H.S. 1993. Increasing CO_2 and plant-plant interactions: effects on natural vegetation. *Vegetatio* 104-105: 157-170.

Johnstone, I.M. 1986. Plant invasion windows: a time based classification of invasion potential. *Biol. Rev.* 61, 369-394.

Jones, W.A. and Sforza, R. 2007. The European Biological Control Laboratory: an existing infrastructure for biological control of weeds in Europe. *OEPP/EPPO Bulletin 37*: 163–165.

Jordon-Thaden, I.E., Louda, SM. 2003. Chemistry of *Cirsium* and *Carduus*: a role in ecological risk assessment for biological control of weeds? *Biochemical Systematics and Ecology* 31: 1353–1396.

Jules, E.S., Kauffman, M.J., Ritts, W.D., Carroll, AL. 2002. Spread of an invasive pathogen over a variable landscape: a nonnative root rot on Port Orford cedar. *Ecology* 83 (11): 3167–3181.

Kaiser, J. 1999. Stemming the tide of invading species. *Science* 285: 1836-1841.

Kanchan, S.D., Jayachandra 1980: Allelopathic effects of *Parthenium hysterophorus* L. IV. Identification of inhibitors. *Plant Soil* 55, 67—75.

Keane, R.M., Crawley, MJ. 2002. Exotic plant invasions and the enemy release hypothesis. *Trends in Ecol. Evol.* 17 (4): 164-170.

Keddy, PA. 1992. Assembly and response rules: Two goals for predictive community ecology. *J. Veg. Sci.* 3: 157-164.

Keeley, JE. 2004. Fire Management Impacts on Invasive Plants in the Western United States. *Conservation Biology* 20(2): 375–384.

Kennedy, TA., Naeem, S., Howe, KM., Knops, JMH., Tilman, D., Reich, P. 2002. Biodiversity as a barrier to ecological invasion. *Nature* 417: 636–638.

Kennedy, T. A. *et al.*, 2002. *Nature* 418:617.

Kerr, JT., Kharouba, HM., Currie, DJ. 2007. Anthropogenic global changes threaten species and the ecosystem services upon which society depends. *Science* 316: 1581-1584.

Khuroo, AA. et al. 2011. Towards an integrated research framework and policy agenda on biological invasions in the developing world: A case-study of India. *Environ. Res.* doi:10.1016/j.envres.2011.02.011.

Kie, JG., Lehmkuhl JF. 2001. Herbivory by wild and domestic ungulates in the Intermountain West. *Northwest Sci.* 75: 55–61.

Kloot, PM. 1987. The invasion of Kangaroo Island by alien plants. *Australian Journal of Ecology* 12: 263–266.

Knops, JMH., Tilman, D., Haddad, N.M., Naeem, S., Mitchell, CE., Haarstad, J., Ritchie ME. Howe KM, Reich PB, Siemann E, Groth, J. 1999. Effects of plant species richness on invasion dynamics, disease outbreaks, insect abundances and diversity. *Ecology Letters* 2: 286–293.

Kohli, R.K., Batish, DR. 1994: Exhibition of allelopathy by *Parthenium hysterophorus* L. in agroecosystems. *Trop. Ecol.* 35: 295—307.

Kolar, C.S., Lodge, D.M. 2001. Progress in invasion biology: predicting invaders. *Trends Ecol. Evol.* 16: 199–204.

Kolar, C.S., Lodge, D.M. 2002. Ecological Predictions and Risk Assessment for Alien Fishes in North America. *Science* 298: 1233.

Kolb, A., Alpert, P., Enters, D, Holzapfel, C. 2002. Patterns of invasion within a grassland community. *Journal of Ecology* 90: 871–881.

Kollmann, J., Bañuelos, MJ. 2004. Latitudinal trends in growth and phenology of the invasive alien plant *Impatiens glandulifera* (Balsaminaceae). *Diversity Distrib.* 10: 377–385.

Kong, C.H., Wang, P., Zhang, C.X., Zhang, M.X., Hu, F. 2006. Herbicidal potential of allelochemicals from *Lantana camara* against *Eichhornia crassipes* and the *alga Microcystis aeruginosa*. *European Weed Research Society* 46: 290–295.

Korhammer, S.A., Haslinger, E. 1994: Isolation of a biologically active substance from rhizomes of quackgrass [*Elymus repens* (L.) Gould]. *J. Agric. Food Chem.* 42: 2048—2050.

Kornissa, G., Caraco, T. 2005. Spatial dynamics of invasion: the geometry of introduced species. *Journal of Theoretical Biology* 233: 137–150.

Krieger, M.J.B., Ross, KG. 2002. Identification of a major gene regulating complex social behavior. *Science* 295: 328–332

Kriticos, D., Brown, J., Radford, I., Nicholas, M. 1999. Plant Population Ecology and Biological Control: Acacia nilotica as a Case Study. *Biological Control* 16: 230–239.

Kriticos, D.J., Sutherst, R.W., Brown, J.R., Adkins, S.W., Maywald, GF. 2003. Climate change and the potential distribution of an invasive alien plant: *Acacia nilotica* ssp. *indica* in Australia. *Journal of Applied Ecology* 40: 111–124.

Kriticos et al. 2003. SPAnDX: a process-based population dynamics model to explore management and climate change impacts on an invasive alien plant, *Acacia nilotica*. *Ecological Modelling* 163: 187–208.

Kull, CA., Tassin, J., Rambeloarisoa, G., and Sarrailh, J.M. 2007. Invasive Australian acacias on western Indian Ocean islands: a historical and ecological perspective. *Afr. J. Ecol.*: 1-6.

Lacey, J.R., Marlow, C.B., Lane, J.R. 1989. Influence of spotted knapweed (*Centaurea maculosa*) on surface runoff and sediment yield. *Weed Technol.* 3: 627–631.

Lass, L.W., Thill, D.C., Shafll, B., Prather, TS. 2002. Remote analysis of biological invasion and biogeochemical change. *Weed Sci.* 16: 426–432.

Lavergne, S., Molofsky, J. 2007. Increased genetic variation and evolutionary potential drive the success of an invasive grass. *Proc. Nati. Acad Sci* 104 (10): 3883–3888.

Lavergne, S., Mouquet, N., Thuiller, W., and Ronce, O. 2010. Biodiversity and Climate Change: Integrating Evolutionary and Ecological Responses of Species and Communities. *Annu. Rev. Ecol. Evol. Syst.* 41,321–50.

Lavorel, S., Prieur-Richard, A.H., Grigulis, K. 1999. Invasibility and diversity of plant communities, from patterns to processes. *Diversity and Distributions* 5: 41–49.

Lee, C.E. 2002. Evolutionary genetics of invasive species. *Trends Ecol. Evol.* 17: 386–391.

Leigh, JH., Briggs, JD. 1992. *Threatened Australian Plants: Overview and Case Studies.* Australian National Parks and Wildlife Service: Canberra.

Leishman, M.R., Haslehurst, T., Ares, A., Baruch, Z. 2007. Leaf trait relationships of native and invasive plants: community- and global-scale comparisons. *New Phytologist* 176: 635–643.

Lemke, D., Hulme, P.E., Browna, J.A., Tadesse, W. 2011. Distribution modelling ofJ apanesehoneysuckle (*Lonicera japonica*) invasion in the Cumber land Plateau and MountainRegion,USA. *Forest Ecology and Management* 262: 139-149

Leppa¨ koski, H.1993. In *Nonindigenous Estuarine and Marine Organisms (NEMO)* (National Oceanic and Atmospheric Administration, Washington, DC, 1993), pp. 37–44; M. A. Ribera and C. F. Boudouresque, in *Progress in Phycological Research*, F. E. Round and D. J. Chapman, Eds. (Biopress, Amsterdam, 1995), vol. 11, pp. 187–268.

Lever, C. 1987. *Naturalized Birds of the World*, Longman

Levine, J.M., D'Antonio, CM.1999. Elton revisited: a review of evidence linking diversity and invasibility. *Diversity and Distrib.* 6: 93–107.

Levine, J.M., Vila, M., D'Antonio, C.M., Dukes, J.S., Grigulis, K., Lavorel, S. 2003. Mechanisms underlying the impacts of exotic plant invasions. *Proceedings of the Royal Society of London* Series B – Biological Sciences 270: 775–781.

Levine, JM. 2000. Species Diversity and Biological Invasions: Relating Local Process to Community Pattern. Science 288: 852-854.

Lewin, R. 1987. Ecological Invasions Offer Opportunities. *Science* 238 (6 November): 752-753.

Li, W.H., Zhang, C., Gao, G., Zan, Q., Yang, Z. 2007. Relationship between *Mikania micrantha* invasion and soil microbial biomass, respiration and functional diversity. *Plant Soil* 296: 197–207.

Li, X., Wilson, SD. 1998. Facilitation among woody plants establishing in an old field. Ecology 79: 2694-2705.

Liddy, J. 1985. A note on the associations of birds and *Lantana* near Beerburrum, south-eastern Queensland. *Corella* 9: 125–126.

Liu-qing, Yu et al. 2007. Response of Exotic Invasive Weed *Alternanthera philoxeroides* to Environmental Factors and Its Competition with Rice. *Rice Science* 14(1): 49-55.

Lockwood, J.L., Cassey, P., and Blackburn, T. 2005. The role of propagule pressure in explaining species invasions. *Trends in Ecol. Evol.* 20 (5): 223-228.

Lodge, DM. 1993. Biological Invasions: lessons for Ecology. *Trends Ecol. Evol.*8(4): 133-137.

Long, J.L. 1981. *Introduced Birds of the World*, Universe Books.

Lonsdale, WM. 1993. Rates of spread of an invading species -*Mimosa pigra* in northern Australia. *J. Ecol.* 81: 513-521.

Lonsdale, WM. 1999. Global patterns of plant invasion and the concept of invasibility. *Ecology* 80: 1522–1536.

Loreau, M., NaeemS, Inchausti P, Bengtsson P, Grime JP, HectorA, Hooper DU, Huston D, Raffaelli MA, Schmid B, Tilman D, Wardle DA. 2001. Biodiversity and Ecosystem Functioning: Current Knowledge and Future Challenges. *Science* 294: 804-808.

Lourenco, P. et al. 2011. Distribution,habitat and biomass of *Pittosporum undulatum*, the most important woody plant invader in the Azores Archipelago. *Forest Ecol. Manage* 262(2): 178-187.

Lovel, GL. 1997. Global change through invasion. *Nature* 388: 627.

Lym, RG. 2005. Integration of biological control agents with other weed management technologies: Successes from the leafy spurge (*Euphorbia esula*) IPM program. *Biological Control* 35: 366–375.

MacDougall, A.S., Boucher, J., Turkington, R. & Bradfield, GE. 2006. Patterns of plant invasion along an environmental stress gradient. *Journal of Vegetation Science* 17: 47-56.

Mack, M.C., D'Antonio, C.M. 2003. Exotic grasses alter controls over soil nitrogen dynamics in a Hawaiian woodland. *Ecol. Appl.* 13 (1): 154–166.

Mack, R.N., Simberloff, D., Lonsdale, W.M., Evans, H., Clout, M., Bazzaz, FA. 2000. Biotic invasions: Causes, epidemiology, global consequences and control. *Ecol. Appl.,* 10: 689–710.

Mack, R.N. 1981. Invasion of *Bromus tectorum* L. into western North America: An ecological chronicle. *Agro-Ecosystems* 7: 145—165.

Mack, RN. 1985. *Studies in Plant Demography*, ed. White, J. (Academic, London), pp. 127–142.

Maestre, FT. 2004. On the importance of patch attributes, environmental factors and past human impacts as determinants of perennial plant species richness and diversity in Mediterranean semiarid steppes. *Diversity Distrib* 10: 21–29.

Maheu-Giroux, M., de Blois, S. 2005. Mapping the invasive species *Phragmites australis* in linear wetland corridors. *Aquatic Botany* 83: 310–320.

Makana, J.R., Thomas, S.C. 2004. Dispersal limits natural recruitment of African mahoganies. *Oikos* 106: 67–72.

Maki, K., Galatowitsch, S. 2004. Movement of invasive aquatic plants into Minnesota (USA) through horticultural trade. *Biological Conservation* 118: 389–396.

Manchester, S.J., Bullock, JM. 2000. The impacts of non-native species on UK biodiversity and the effectiveness of control. *Journal of Applied Ecology* 37: 845-864.

Mandak, B. 2003. Germination requirements of invasive and non-invasive *Atriplex* species: a comparative study. *Flora* 198: 45–54.

Mangla, S., Inderjit & Callaway, R.M. 2008. Exotic invasive plant accumulates native soil pathogens which inhibit native plants. *Journal of Ecology* 96: 58–67.

Marchante E, Kjøller A, Struwe S, Freitas H. 2008. Short- and long-term impacts of *Acacia longifolia* invasion on the belowground processes of a Mediterranean coastal dune ecosystem. *Applied Soil Ecology* 40 (2): 210-217.

Markin, G.P., Lai, P.Y., Funasaki, GY. 1992. Status of biological control of weeds in Hawaii and implications for managing native ecosystems. *In* "*Alien Plant Invasions in Native Ecosystems of Hawaii: Management and Research*" (C. P. Stone, C.W. Smith, and J. T. Tunison, Eds.), Univ. of Hawaii Press, Honolulu. pp. 466–482.

Marshall, V.J. 2000. Impacts of forest harvesting on biological processes in northern forest soils. *Forest Ecology and Management* 133: 43-60.

Marx J. 2004. The Roots of Plant-Microbe Collaborations. *Science* 304: 234-236.

Mascaro, J., Becklund, K.K., Hughes, R.F., Schnitzer, SA. 2008. Limited native plant regeneration in novel, exotic-dominated forests on Hawai'i. *Forest Ecology and Management* 256: 593–606.

Maskell, L.C., Firbank, L.G., Thompson, K., Bullock, J.M., Smart, SM. 2006. Interactions between non-native plant species and the floristic composition of common habitats. *Journal of Ecology* 94: 1052–1060.

Mason, T.J., French, K. 2007. Management regimes for a plant invader differentially impact resident communities. *Biological Conservation* 136: 2 4 6 –2 5 9.

Mauchamp, A. 1997.Threats from alien plant species in Galapogas islands. *Conservation Biology* 11(1): 260-263.

McCaughey, T.L., Stephenson, G.R. 2000. Time from flowering to seed viability in purple loosestrife (*Lythrum salicaria*). *Aquatic Botany* 66: 57–68.

McConnachie, A.J., de Wit, M.P., Hill, M.P., Byrne, M.J. 2003. Economic evaluation of the successful biological control of *Azolla filiculoides* in South Africa. *Biological Control* 28: 25–32

McDougall, K.L., Morganb, J.W., Walshc, N.G., Williams, R.J. 2005. Plant invasions in treeless vegetation of the Australian Alps. Perspectives in Plant Ecology, *Evolution and Systematics* 7: 159–171.

Mcgeoch, M.A., Chown, S.L., Kalwija, J.M. 2006. Global Indicator for Biological Invasion. *Conservation Biology* 20 (6): 1635–1646.

McGrady-Steed, J., Harris, P., Morin P. 1997. Biodiversity regulates ecosystem reliability. *Nature* 390, 162–165.

McKnight, B.N., editor. 1993. *Biological Pollution: The Control and Impact of Invasive Exotic Species*. Indiana Academy Press, Indianapolis.

McNeely, J.A., Kapoor-Vijay, P., Zhi, L., Olsvig-Whittaker, L., Sheikh, K.M., Smith AT. 2009. Conservation biology in Asia: The major policy challenges. *Conserv. Biol.* 23: 805–810.

McNeely, JA. 2001. *An introduction to human dimensions of invasive alien species. Human dimension of the consequences of invasive alien species.* ISSG, IUCN, www.issg.org.

Meiman, P.J., Redente, E.F., Paschke, M.W. 2006. The role of the native soil community in the invasion ecology of spotted (*Centaurea maculosa* auct. non Lam.) and diffuse (*Centaurea diffusa* Lam.) knapweed. *Applied Soil Ecology* 32: 77–88.

Meyer, J.Y., Lavergne, C. 2004. *Beautés fatales* : Acanthaceae species as invasive alien plants on tropical Indo-Pacific Islands. Diversity and Distributions 10: 333–347.

Milberg, P., Lamont, BB. 1995. Fire enhances weed invasion of roadside vegetation in southwestern Australia. Biological Conservation 73: 45–49.

Milchunas, D.G., & Lauenroth, W.K. 1995 Inertia in plant community structure: state changes after cessation of nutrient- enrichment stress. *Ecological Applications* 5: 452–458.

Mohan, J.E., Ziska, L.H., Schlesinger, W.H., Thomas, R.B., Sicher, R.C., George, K., Clark, JS. 2006. Biomass and toxicity responses of poison ivy (*Toxicodendron radicans*) to elevated atmospheric CO_2. *Proc. Nati. Acad Sci* 13 (103: No. 24), 9086–9089.

Mollo, et al. 2008.Factors promoting marine invasions: A chemoecological approach. *Proc. Nati. Acad Sci* 105 (12): 4582–4586.

Mooney, H., Cropper, A., Reid, A. 2005. Confronting the human dilemma. *Nature* 434: 561-562.

Mooney, H.A., Cleland, EE. 2001. The evolutionary impact of invasive species. *Proc. Nati. Acad Sci* 98 (10): 5446–5451.

Mooney, H.A., Hobbs R.J. 2000. *Invasive Species in a Changing World* (Island Press, Washington, DC).

Mooney, H.A. 1999. Species without frontiers. *Nature* 397: 665–666.

Morghan, K.J.R., Seastedt 1999. Effects of soil Nitrogen reduction reduction on non-native plants in restored grasslands. *Restoration Ecology* 7(1): 51-55.

Mueller-Dombois, D. 2000. Rain forest establishment and succession in the Hawaiian Islands. *Landscape and Urban Planning* 51:147-157.

Murcia, C. 1995. Edge effects in fragmented forests: implications for conservation. *Trends Ecol. Syst.* 10: 58–62.

Myers, J.H., Simberloff, D., Kuris, A.M., Carey, JR. 2000. Eradication revisited: dealing with exotic species. *Trends in Ecol. Evol.*15 (8): 316-320.

Myster, R.W. 2012. Plants Replacing Plants: The Future of Community Modeling and Research. *Bot. Rev.* 78:2–9

Naeem, S., Li, S.1997. Biodiversity enhances ecosystem reliability. *Nature* 390: 507–509

Naeem, S., et al. 2000. Plant diversity increases resistance to invasion in the absence of covarying extrinsic factors. *Oikos* 91: 97–108.

Naiman, R., Décamps, H. 1997. The ecology of interfaces: riparian zones. *Annual Review of Ecology and Systematics,* 28: 621–658.

Nalepa, T.F., and Schloesser, DW. Eds., 1993. *Zebra Mussels: Biology, Impacts, and Controls* (Lewis, Boca Raton, FL, 1993); J. Travis, *Science* 262: 1366.

Neilan, W., Catterall, C.P., Kanowski, J., Stephen, McKenna. 2006. Do frugivorous birds assist rainforest succession in weed dominated oldfield regrowth of subtropical Australia? *Biological Conservation* 129: 393-407.

Nernberg, D., & Dale, M.R.T. 1997 Competition of five native prairie grasses with *Bromus inermis* under three moisture regimes. *Canadian Journal of Botany* 75: 2140–2145.

Newell, GR. 1998. Characterization of vegetation in an Australian open forest community affected by cinnamon fungus (*Phytophthora cinnamomi*): implications for faunal habitat quality. *Plant Ecol.* 137 (1): 55–70.

Nunez, M.A., Pauchard, A. 2010. Biological invasions in developing and developed world: does one model fit all? *Biol. Invasions* 12: 707–714.

Osunkoya, O.O., Perrett, C. 2011. *Lantana camara* L. (Verbenaceae) invasion effects on soil physicochemical properties. *Biol Fertil Soils* 47:349–355.

Palmer, T.M., Stanton, M.L., Young, T.P., Goheen, J.R., Pringle, R.M., Karban, R. 2008. Breakdown of an Ant-Plant Mutualism Follows the Loss of Large Herbivores from an African Savanna. *Science* 319: 192-195.

Palumbi, S. R. 2001. Humans as the world's greatest evolutionary force. *Science* 293: 1786-1790.

Pandey, DK. 1994. Inhibition of salvinia (*Salvinia molesta* Mitchell) by parthenium (*Parthenium hysterophorus* L.). II. Relative effect of flower, leaf, stem, and root residue on salvinia and paddy. *J. Chem. Ecol.* 20: 3123—3131.

Parker, I.M. et al. 1999. Impact: towards a framework for understanding the ecological effects of invaders. *Biol. Inv.* 1: 3–19.

Parker, I.M. Kareiva, P. 1996. Assessing the risk of invasion for genetically engineered plants: Acceptable evidence and reasonable doubt. *Biological Conservation* 78: 193-203.

Parker, J.D., Burkepile, D.E., Hay, ME. 2006. Opposing Effects of Native and Exotic Herbivores on Plant Invasions. *Science* 311: 1459-1461.

Pauchard, A., Cavieres, LA., Bustamante, R.O. 2004. Comparing alien plant invasions among regions with similar climates: where to from here? *Diversity and Distributions* 10: 371–375.

Pe'tillon, J., Ysnel, F., Canard, A., Lefeuvre, J.C. 2005. Impact of an invasive plant (*Elymus athericus*) on the conservation value of tidal salt marshes in western France and implications for management: Responses of spider populations. *Biological Conservation* 126: 103–117.

Pearson, D.E. and Callaway, R.M. 2003. Indirect effects of host-specific biological control agents. *Trends in Ecol. Evol.* 18(9), 456-461.

Peart, D.R., Foin, T.C. 1985. Analysis and prediction of population and community change: a grassland case study. *The Population Structure of Vegetation* (ed. J. White), pp. 313–339. Junk, Dordrecht.

Peterson, A.T., Vieglais, D.A. 2001. Predicting species invasions using ecological niche modeling: new approaches from bioinformatics attack a pressing problem. *Bioscience* 51: 363–371.

Peterson, AT. 2003. Predicting the geography of species invasions via ecological niche modeling. *Q. Rev. Biol.* 78: 419–433.

Petit, R.J. 2004. Biological invasions at the gene level. *Diversity Distrib.* 10:, 159–165.

Pfisterer, A.B., Joshi, J., Schmid, B., Fischer, M. 2004. Rapid decay of diversity-productivity relationships after invasion of experimental plant communities. *Basic Appl. Ecol.* 5, 5–14.

Pickard, J. 1984. Exotic plants on Lord Howe Island: distribution in space and time, 1853–1981. *Journal of Biogeography* 11: 181–208.

Pimentel, D., Lach, L., Zuniga, R., Morrison, D. 2000. Environmental and Economic Costs of Nonindigenous Species in the United States. *Bioscience* 50 (1): 53-65.

Pimentel et al. 2001. Economic and environmental threats of alien plant, animal, and microbe invasions. *Agriculture, Ecosystems and Environment* 84: 1–20.

Pimm, S.L. 2007. Africa: Still the "Dark Continent". *Conservation Biology* 21 (3): 567–569.

Pimm, S.L., Russell, G.J., Gittleman, J.L., Brooks, TM. 1995. The future of biodiversity. *Science* 269: 347–350.

Postle, A.C., Majer, J.D., Bell, DT. 1986. Litter invertebrates and litter decomposition in Jarrah (Eucalyptus marginata) forest affected by Jarrah dieback fungus (*Phytophthora cinnamoni*). *Pedobiologia* 29 (1): 47–69.

Powell, RD. 1990. The role of spatial pattern in the population biology of *Centaurea diffusa*. *Journal of Ecology* 78, 374–388.

Pugh, PJA. 1994. Non-indigenous acari of Antarctica and the sub-Antarctic islands. *Zoological Journal of the Linnean Society* 110: 207–217.

Purvis, A., Hector, A. 2000. Getting the measure of biodiversity. *Nature* 405: 212-219.

Pysek, P., Prach, K., Rejmánek, M., Wade, PM. 1995. *Plant Invasions: General Aspects and Special Problems*. SPB Academic Publishing, Amsterdam.

Pysek, P., Richardson, D.M., Rejmánek, M., Webster, G.L., Williamson, M., Kirschner, J. 2004. Alien plants in checklists and floras: towards better communication between taxonomists and ecologists. *Taxon* 53: 131–143.

Rai, P.K. 2013. Forest and land use mapping using Remote Sensing & Geographical Information System: A case study on model system. *nvironmental Skeptics and Critics;* 2(3): In Press.

Rai, P.K. and Rai, P.K. 2013. Paradigms of global climate change and sustainable development: Issues and related policies. *Environmental Skeptics and Critics,* 2(2): 30-45.

Rai, P.K. 2012. Assessment of Multifaceted Environmental Issues and Model Development of an Indo- Burma Hot Spot Region. *Environmental Monitoring & Assessment* 184:113–131.

Rai, P.K. and Lalramnghinglova, H. 2011a. Ethnomedicinal plants of India with special reference to an Indo-Burma hotspot region: An overview. *Ethnobotany Research & Apllications;* 9: 379-420.

Rai, P.K. and Lalramnghinglova, H. 2010a. Ethnomedicinal Plants from Agroforestry Systems and Home gardens of Mizoram, North East India. *Herba Polonica* 56 (2): 1-13.

Rai, P.K. 2011. Land Use Changes in North Eastern Himalayan Region (an Indo-Burma Hot spot) and its Impact on Human Health" *In Biodiversity And Sustainable Development* (Ed. K.N. Tiwari and S. Lata), Prasanna Prakashan, Bhopal, 175-192.

Rai, P.K. and Lalramnghinglova, H. 2011b.Threatened and less known ethnomedicinal plants of an Indo-Burma hotspot region: conservation implications. *Environmental Monitoring and Assessment* 178: 53–62.

Rai, P.K. and Lalramnghinglova, H. 2010b. Lesser known ethnomedicinal plants of Mizoram, North East India: An Indo-Burma hotspot region. *Journal of Medicinal Plants Research* 4(13): 1301-1307.

Rai, P.K. and Lalramnghinglova, H. 2010c. Ethnomedicinal Plant Resources of Mizoram, India: Implication of Traditional Knowledge in Health Care System. *Ethnobotanical Leaflets* 14: 274-305.

Rai, P.K. 2009. Comparative Assessment of Soil Properties after Bamboo Flowering and Death in a Tropical Forest of Indo-Burma Hot spot. *Ambio: A Journal on Human Environment* 38 (2): 118-120.

Radford, I.J., Dickinson, K.J.M., Lord, J.M. 2006. Nutrient stress and performance of invasive *Hieracium lepidulum* and co-occurring species in New Zealand. *Basic and Applied Ecology* 7: 20-333.

Raffaelli, D. 2004. How Extinction Patterns Affect Ecosystems. *Science* 306: 1141-1142.

Raghu et al. 2006. Adding biofuels to the invasive species fire? *Science* 313: 1742.

Rahel, FJ. 2000. Homogenization of fish faunas across the United States. *Science* 288: 854–856.

Ramakrishnan. P.S. (ed.) 1991. *Ecology of Biological Invasion in the Tropics*, International Scientific Publications, New Delhi, 1991.

Rambuda, T.D., Johnson, S.D. 2004. Breeding systems of invasive alien plants in South Africa: does Baker's rule apply? *Diversity and Distributions:* 10: 409–416.

Rapoport, EH. 1991 Tropical versus temperate weeds: A glance into the present and future. In *Ecology of Biological Invasion in the Tropics* (ed. Ramakrishnan, P. S.), International Scientific Publications, New Delhi pp. 441–451.

Rapson, G.L., Wilson, JB. 1992. Genecology of *Agrostis capillaries* L. (Poaceae) — an invader into New Zealand. 1. Floral phenology. *New Zealand Journal of Botany* 30: 1–11.

Raven, P.H., Johnson, GB. 1992. *Biology,* 3rd Edition. Mosby Year Book, St. Louis, MO.

Rees M, Condit R, Crawley M, Pacala S, Tilman D. 2001. Long-Term Studies of Vegetation Dynamics. *Science* 293: 350-355.

Regal, PG. 1977. Ecology and Evolution of Flowering Plant Dominance. *Science* 196: 622-629.

Reinhart, K.O., Greene, E., and Callaway, R.M. 2005. Effects of *Acer platanoides* invasion on understory plant communities and tree regeneration in the northern Rocky Mountains. *Ecography* 28: 573-582.

Rejmanek, M., Randall, J.M. 1993, Invasive alien plants in California: Summary and comparison with other areas in North America. *Madrono* 41: 161–177.

Rejmánek, M. 1999. Holocene invasions: finally the resolution ecologists were waiting for! *Trends in Ecol. Evol.* 14 (1): 18-20.

Reynolds, H.L., Packer, A., Bever, J.D., Clay, K. 2003. Grassroots ecology: plant–microbe–soil interactions as drivers of plant community structure and dynamics. *Ecology* 84: 2281–2291.

Rice, E.L. 1974. *Allelopathy.* Academic Press, New York.

Rice, K.J., Mack, R.N. 1991. Ecological genetics of *Bromus tectorum. III.* The demography of reciprocally sown populations. *Oecologia* 88: 91–101.

Richardson, D. M. and Rejmánek, M. 2004. Conifers as invasive aliens: a global survey and predictive framework. *Diversity and Distributions* 10: 321–331.

Richardson, D.M., Blanchard, R.2011. Learning from our mistakes:minimizing problems with invasive biofuel plants. *Curr. Opin. Environ.Sus.* 3, 36–42.

Richardson DM, Cowling RM. 1992. *Why is mountain fynbos invasible and which species invade?* In: B.W. van Wilgen, D.M. Richardson, F.J. Kruger and H.J. van Hensbergen (Editors), *Fire in South African Mountain Fynbos.* Springer-Verlag, Berlin, pp. 161-181.

Richardson, D.M., Holmes, P.M., Esler KJ, Galatowitsch SM, Stromberg JC, Kirkman SP, Pyšek P, Hobbs RJ. 2007. Riparian zones — degradation, alien plant invasions and restoration prospects. *Diversity and Distributions* 13: 126–139.

Richardson, D.M., Macdonald, I.A.W., Hoffmann, J.H., Henderson, L. 1997. Alien plant invasions. In: *Vegetation of Southern Africa* (ed. by R.M. Cowling, D.M. Richardson and S.M. Pierce), pp. 535–570. Cambridge University Press.

Richardson, D.M., Macdonald, I.A.W., Holmes, P.M., Cowling, R.M. 1992. Plant and animal invasions. In: R.M. Cowling (Editor), *The Ecology of the Fynbos.* Oxford University Press, Cape Town, pp. 271-308.

Richardson, D.M. et al. 2000. Naturalization and invasion of alien plants – concepts and definitions. *Div. Distrib.* 6: 93–108.

Richardson, D.M., Pysek, P., Rejmánek, M., Barbour, M.G., Panetta, F.D. & West, C.J. 2000. Naturalization and invasion of alien plants: concepts and definitions. *Diversity and Distributions.*

Richardson, D.M.(ed.) 2011. *Fifty Years of Invasion Ecology: The Legacy of Charles Elton*, Wiley-Blackwell

Riggs, R.A., Cook, J.G., Irwin, L.L. 2005. Management implications of ungulate herbivory in Northwest forest ecosystems. In:Wisdom, M.J. (tech. Ed.), The Starkey Project: *A Synthesis of Long-Term Studies of Elk and Mule Deer.* Alliance Communications Group, Lawrence, Kansas, pp. 217–232.

Riggs, R.A., Tiedemann, A.R., Cook, J.G., Ballard, T.M., Edgerton, P.J., Vavra, M., Krueger, W.C., Hall, F.C., Bryant, L.D., Irwin, L.L., DelCurto, T. 2000. *Modification of mixed-conifer forests by ruminant herbivores in the Blue Mountains Ecological Province.* USDA, Forest Service, Pacific Northwest Station, Research Paper PNW-RP-527.

Robbins, P. 2001. Tracking invasive land covers in India, or why our landscapes have never been modern. *Annals of Association of American Geographers* 91 (4), 637-659.

Robertson, M.P., Villet, M.H., Palmer, A.R. 2004. A fuzzy classification technique for predicting species' distributions: applications using invasive alien plants and indigenous insects. *Diversity and Distributions* 10: 461–474.

Robinson, G.R., Quinn, J.F. & Stanton, ML. 1995 Invasibility of experimental habitat islands in a California winter annual grassland. *Ecology* 76: 786–794.

Roche, B.F. 1994. *Status of knapweeds in Washington.* Washington State University Cooperative Extension Service, Knapweed Newsletter 8, 2-4.

Rodiyati, A., Arisoesilaningsih, E., Isagi, Y., Nakagoshi, N. 2005. Responses of *Cyperus brevifolius* (Rottb.) Hassk. and *Cyperus kyllingia* Endl. to varying soil water availability. *Environmental and Experimental Botany* 53: 259–269.

Root, T.L., Price, J.T., Hall, K.R., Schneider, S.H., Rosenzweig, C., Pounds, A. 2003. Fingerprints of global warming on wild animals and plants. *Nature* 421: 57–60.

Rosenzweig, ML. 2001 The four questions: what does the introduction of exotic species do to diversity? *Evol. Ecol. Res.* 3: 361–367.

Rouget, M., Richardson, D.M., Nel, J.L., Le Maitre, D.C., Egoh, B., Mgidi, T. 2004. Mapping the potential ranges of major plant invaders in South Africa, Lesotho and Swaziland using climatic suitability. *Diversity and Distrib.*10, 475–484.

Rougeta, M., Richardson, D.M., Cowling, R.M., Lloyd, J.W., Lombard, AT. 2003. Current patterns of habitat transformation and future threats to biodiversity in terrestrial ecosystems of the Cape Floristic Region, South Africa. *Biological Conservation* 112, 63–85.

Rozefelds, A.C.F., Cave, L., Morris, D.I. & Buchana, A.M. 1999. The weed invasion in Tasmania since 1970. *Australian Journal of Botany* 47: 23–48.

Sa´ nchez-Floresa, E., Rodrı´guez-Gallegosb, H., Yool, S.R. 2008. Plant invasions in dynamic desert landscapes. A field and remote sensing assessment of predictive and change modelling. *Journal of Arid Environments* 72: 189–206.

Sakai, et al. 2001. The population biology of invasive species. *Annu. Rev. Ecol. Syst.* 32,305–32.

Sala, O.E., Chapin, F.S. III, Armesto, J.J., Berlow, E., Bloomfield, J., Dirzo, R., Huber-Sanwald, E., Huenneke, L.F., Jackson, R.B., Kinzig, A., Leemans, R., Lodge, D.M., Mooney, H.A., Oesterheld, M., Poff, N.L., Sykes, M.T., Walker, B.H., Walker, M., Wall, D.H. 2000. Global biodiversity scenarios for the year 2100. *Science* 287: 1770–1774.

Samways, M.J., Caldwell, PM., Osborn, R. 1996. Ground-living invertebrate assemblages in native, planted and invasive vegetation in South Africa. *Agriculture, Ecosystems and Environment* 59: 19–32.

Sanders, N.J., Gotelli, N.J., Heller, N.E., Gordon, D.M. 2003. Community disassembly by an invasive species *Proc. Nati. Acad Sci* 100 (5): 2474-2477.

Sasek, T.W. and Strain, BR. 1991. Effects of CO_2 enrichment on the growth and morphology of a native and an introduced honeysuckle vine. *Am. J. Bot.* 78: 69–75

Sasek, T.W., Strain, BR. 1988. Effects of carbon dioxide enrichment on the growth and morphology of kudzu (*Pueraria lobata*), *Weed Sci.* 36: 28–36

Saunders, S.C., Mislivets, M.R., Chen, J., Cleland, DT. 2002. Effects of roads on landscape structure within nested ecological units of the Northern Great Lakes Region, USA. *Biological Conservation* 103, 209–225.

Sax, DF. 2001. Latitudinal gradients and geographic ranges of exotic species: implications for biogeography. *Journal of Biogeography* 28: 139-150.

Sax, D. F. Stachowicz, J.J. and Gaines, S.D. (Eds.) 2005. *Species Invasions: Insights into Ecology, Evolution, and Biogeography* Sinauer, Sunderland, MA, 509 pp.

Saxena, MK. 2000. Aqueous leachate of *Lantana camara* kills water hyacinth. *J. Chem. Ecol.* 26: 2435—2447.

Scheffer et al. 2003. Floating plant dominance as a stable state. *Proc. Nati. Acad Sci 100* (7): 4040-4045.

Schei, PJ. 1996. Conclusions and recommendations from the UN/ Norway conference on alien species. *Sci. Int.* 63: 32–36.

Scheiner, SM. 1993 Genetics and evolution of phenotypic plasticity. *Annu. Rev. Ecol. Syst.* 24:35–68.

Schlichting, C.D., Pigliucci, M. 1998. *Phenotypic Evolution: A Reaction Norm Perspective*. Sinauer, Sunderland, Massachusetts.

Schlichting, CD. 1986. The evolution of phenotypic plasticity in plants. *Annual Review of Ecology and Systematics,* 17: 667–693.

Schwartz, MW. et al. 2000. Linking biodiversity to ecosystem functioning: implications for conservation ecology. Oecologia 122: 297–305.

Scott, J.K. 2001. Europe gears-up to fight invasive species (meeting report). Trends in Ecol. Evol. 16 (4): 171-172.

Seastedt, T. 2009. Traits of plant invaders. *Nature* 459: 783-784.

Serbesoff-King, K. 2003. *Melaleuca* in Florida: a literature review on the taxonomy, distribution, biology, ecology, economic importance and control measures. *Journal of aquatic plant management* 41: 98–112.

Shabbir, A., Bajwa, R. 2006. Distribution of parthenium weed (*Parthenium hysterophorus*L.), an alien invasive weed species threatening the biodiversity of Islamabad. *Weed Biology and Management* 6: 89–95.

Sharma, G.P., Raghubanshi, A.S., Singh, JS. 2005. *Lantana* invasion: An overview. *Weed Biology and Management* 5: 157–165.

Sharma, G.P., Raghubanshi, AS. 2010. How *Lantana* invaded India. Curr. *Conserv.* 4 (1): 21–22.

Sharma, G.P., Raghubanshi, AS. 2009. *Lantana* invasion alters the soil pools and processes: a case study with special reference to nitrogen dynamics in the tropical dry deciduous forest of India. *Appl. Soil Ecol.* 42, 134–140.

Sharma, G.P., Singh, J.S., Raghubanshi, AS. 2005. Plant invasions: Emerging trends and future implications. Current Science 88(5): 726-734.

Sharma, R., Gupta, R. 2007. *Cyperus rotundus* extract inhibits acetylcholinesterase activity from animal and plants as well as inhibits germination and seedling growth in wheat and tomato. Life Sciences 80: 2389–2392.

Shea, K. and Chesson, P. 2002. Community ecology theory as a framework for biological invasions. *Trends in Ecol. Evol.* 17(4), 170-176.

Sheil, D. 2001. Conservation and Biodiversity monitoring in tropics: Realities, priorities and Distractions. *Conservation Biology* 15 (4): 1179-1182.

Sheley, R.L., Jacobs, J.S., Carpinelli, M.F. 1998. Distribution, biology and management of diffuse knapweed (*Centaurea diffusa*) and spotted knapweed (*Centaurea maculosa*).*Weed Technol.* 12: 353– 362.

Sheppard, A.W., Shaw, R.H., Sforza, R. 2006. Top 20 environmental weeds for classical biological control in Europe: a review of opportunities, regulations and other barriers to adoption. *Weed Research* 46: 93–117.

Sherry, R.A., Zhou, X., Gu, S., Arnone,J.A., Schimel, D.S., Verburg, P.S., Wallace, L.L., Luo, Y. 2007. Divergence of reproductive phenology under climate warming. *Proc. Nati. Acad Sci* 104 (1) :198-202.

Shiferaw, H., Teketay., D., Nemomissa, S., Assefa, F. 2004. Some biological characteristics that foster the invasion of *Prosopis juliflora* (Sw.) DC. at Middle Awash Rift Valley Area, north-eastern Ethiopia. *Journal of Arid Environments* 58: 135–154.

Shukla, J., Nobre, C., Sellers, P. 1990. Amazon deforestation and climate change. S*cience* 247, 1322–1325.

Siemann, E., Rogers, W.E., Grace, JB. 2007. Effects of nutrient loading and extreme rainfall events on coastal tallgrass prairies: invasion intensity, vegetation responses, and carbon and nitrogen distribution. *Global Change Biology* 13, 2184–2192.

Simberloff, D. 2000. Global climate change and introduced species in United States forests. *Science of the Total Environment* 262: 253–261.

Simberloff, D. 2001.Eradication of island invasives: practical actions and results achieved. *Trends in Ecol. Evol.* 16 (6): 273-274.

Simberloff, D. 2004. Review of "Invasion biology: critique of a pseudoscience". *Ecol. Econ.* 48: 360–362.

Singh, HP., Batish, D.R., Pandher, J.K,. Kohli, RK. 2003a. Assessment of allelopathic properties of *Parthenium hysterophorus* residues. *Agric. Ecosyst. Environ.* 95: 537—541.

Slobodchikoff, C. N., Doyen, J. T. 1997. Effects of *Ammophila arenaria* on sand dune arthropod communities. Ecology 58: 1171–1175.

Slobodkin, LB. 2001. The good, the bad and the reified. *Evol. Ecol. Res.* 3: 1–13.

Smith, H., Firbank, L.G., Macdonald, D.W. 1999. Uncropped edges of arable fields managed for biodiversity do not increase weed occurrence in adjacent crops. *Biological Conservation* 89: 107-111.

Smith, J.M.D., Warda, J.P., Child, L.E., Owen, MR. 2007. A simulation model of rhizome networks for *Fallopia japonica* (Japanese knotweed) in the United Kingdom. *Ecological modelling* 200: 421–432.

Smith, RIL. 1996. Introduced plants in Antarctica : potential impacts and conservation issues. *Biological Conservation* 76: 135–146.

Smith, S.D., Strain, B.R. Sharkey, T.D. 1987. Effects of carbon dioxide enrichment on four Great Basin [USA] grasses, *Funct. Ecol.* 1: 139–144

Sousa, WP. 1984. The role of disturbance in natural communities. *Annu. Rev. Ecol. Syst.* 15, 353–391.

Stachon, W.J., Zimdahl, R.L. 1980. Allelopathic activity of Canada thistle (*Circium arvense*) in Colorado. *Weed Sci.* 28: 83—86.

Stachowicz, J.J., Terwin J.R., Whitlatch, R.B., Osman, R.W. 2002. Linking climate change and biological invasions: Ocean warming facilitates nonindigenous species invasions. *Proc. Nati. Acad Sci* 99 (24): 15497–15500.

Stachowicz, J.J., Whitlatch, R.B., Osman, R.W. 1999. Species diversity and invasion resistance in a marine ecosystem. *Science* 286: 1577–1579.

Stadler, J., Trefflich, A., Klotz, S., & Brandl, R. 2000. Exotic plant species invade diversity hot spots: alien flora of north-western Kenya. *Ecography,* 23: 169–176.

Stampe, E.D., and Daehler, C.C. 2003. Mycorrhizal species identity affects plant community structure and invasion: a microcosm study. *Oikos* 100: 362–372.

Stastny, M., Schaffner, U.R.S., Elle, E. 2005.Do vigour of introduced populations and escape from specialist herbivores contribute to invasiveness? *Journal of Ecology* 93: 27–37.

Stebbins, G.L. 1957. Self fertilization and population variability in the higher plants. *American Naturalist*, 91: 337–354.

Stermitz,F.R., Bais, H.P., Foderaro, T.A., Vivanco, JM. 2003. 7,8-Benzoflavone: a phytotoxin from root exudates of invasive Russian knapweed. *Phytochemistry* 64: 493–497

Stohlgren, T.J., Binkley, D., Chong, G.W., Kalkhan, M.A., Schell, L.D., Bull, K.A., Otsuki, Y., Newman, G., Bashkin, M., Son, Y. 1999. Exotic plant species invade hot spots of native plant diversity. Ecological Monographs, 69: 25–46.

Strauss, S.Y., Webb, C.O., Salamin, N. 2006. Exotic taxa less related to native species are more invasive. *Proc. Nati. Acad Sci* 103 (15): 5841–5845.

Sumners, W.H., Archibold, OW. 2007. Exotic plant species in the southern boreal forest of Saskatchewan. *Forest Ecology and Management* 251: 156–163.

Sutherst, R.W., Maywald, G.F., Russell, BL. 2000. Estimating vulnerability under global change: modular modelling of pests. *Agric. Ecosys. Environ.* 82, 303–319.

Sutherst, R.W., Maywald, GF. 1991. Climate modelling and pest establishment climate-matching for quarantine using Climex. *Plant Prot. Q.* 6: 3–7.

Swaminathan, MS. 2003. Bio-diversity: an effective safety net against environmental pollution. *Environmental Pollution* 126: 287–291.

Sweeney, B.W., Czapka, SJ. 2004. Riparian forest restoration: why each site needs an ecological prescription. *Forest Ecology and Management* 192: 361–373.

Symstad, AJ. 2000. A test of the effects of functional group richness and composition on grassland invasibility. *Ecology* 81: 99–109 (2000).

Taylor, B.W., Irwin, RE. 2004. Linking economic activities to the distribution of exotic plants. Proc. Nati. Acad Sci 101 (51): 17725–17730.

Taylor, K., Potvin, C. 1997. Understanding the long-term effect of CO_2 enrichment on a pasture: the importance of disturbance. *Can. J. Bot.* 75: 1621–1627

Tefera, T. 2002. Allelopathic effects of *Parthenium hysterophorus* extracts on seed germination and seedling growth of *Eragrostis* tef. *J. Agron. Crop Sci.* 188: 306—310.

Theoharides, K.A. Dukes, J.S. 2007. Plant invasion across space and time: factors affecting non-indigenous species success during four stages of invasion. *New Phytologist* 176: 256–273.

Thibault, K.M. Brown, J.H. 2007. Impact of an extreme climatic event on community assembly. *Proc. Nati. Acad Sci* 105 (9): 3410–3415.

Thompson, K., Hodgson, J.G., Grime, J.P., Burke, M.J.W. 2001. Plant traits and temporal scale: evidence from a 5-year invasion experiment using native species. Journal of Ecology 89: 1054–1060.

Thomson, A.G., Radford, G.L., Norris, D.A., Good, J.E.G., 1993. Factors affecting the distribution and spread of *Rhododendron* in north Wales. *J. Environ. Manage.,* 39: 199-212.

Thornby, D., Spencer, D., Hanan, J. and Sher, A. 2007. L-DONAX, a growth model of the invasive weed species, *Arundo donax* L. *Aquatic Botany* 87: 275–284.

Tilman, D. 1993. Species richness of experimental productivity gradients, how important is colonization limitation. *Ecology* 74: 2179–2191.

Tilman, D. 1997. Community invasibility, recruitment limitation, and grassland biodiversity. *Ecology* 78: 81–92.

Tilman, D. 1999. The ecological consequences of changes in biodiversity: A search for general principles. *Ecology* 80: 1455–1474.

Tilman, D. and Lehman, C. 2001. Human-caused environmental change: Impacts on plant diversity and evolution. *Proc. Nati. Acad Sci* 98 (10): 5433–5440.

Tilman,D., Fargione, J., Wolff, B., Carla, D.Õ., Antonio, Dobson, A., Howarth,R.,Schindler, D., Schlesinger,W.H., Simberloff, D., Swackhamer, D. 2001. Forecasting Agriculturally Driven Global Environmental Change. *Science* 292: 281

Timsina, B., Shrestha, B.B., Rokaya, M.B., Münzbergová, Z. 2011. Impact of *Parthenium hysterophorus L.* invasion on plants pecies composition and soil properties of grassland communities in Nepal. *Flora* 206: 233–240.

Totland, O., Nyeko, P., Bjerknes, A., Hegland, S.J., Nielsen, A. 2005. Does forest gap size affects population size, plant size, reproductive success and pollinator visitation in *Lantana camara*, a tropical invasive shrub? *Forest Ecology and Management* 215: 329–338.

Tracya, B.F., Renne, I.J., Jim, Gerrish J., Sanderson, MA. 2004. Effects of plant diversity on invasion of weed species in experimental pasture communities. *Basic and Applied Ecology* 5: 543-550.

Trakhtenbrot, A., Nathan, R., Perry, G., Richardson, D.M. 2005. The importance of long-distance dispersal in biodiversity conservation. *Diversity Distrib.* 11: 173–181.

Traveset, A., Richardson, DM. 2006. Biological invasions as disruptors of plant reproductive mutualisms. *Trends in Ecol. Evol.* 21 (4): 208-216.

Trethowan, P.D., Robertsonb, M.P., McConnachie, AJ. 2011. Ecological niche modelling of an invasive alien plant and its potential biological control agents. *South African Journal of Botany* 77: 137–146.

Tripathi, R.S., Singh, R.S., Rai, JPN. 1981. Allelopathic potential of *Eupatorium adenophorum* – A dominant ruderal weed of Meghalaya. *Proc. Indian. Natl. Sci. Acad. Part* B 47: 458– 465.

Tsutsui, N.D. et al. 2000. Reduced genetic variation and the success of an invasive species. *Proc. Natl. Acad. Sci.* 97: 5948–5953

U.S. Congress, Office of Technology Assessment, 1993. *Harmful Non-Indigenous Species in the United States,* OTA-F-565 (Washington, DC: U.S. Government Printing Office, September 1993).

Underwood, E., Ustin, S. DiPietro, D. 2003. Mapping non-native plants using hyperspectral imagery. *Remote Sensing Environ.* 86: 150–161.

USDA, NRCS, 2002. *The PLANTS Database*, Version 3.5, http://plants.usda.gov. National Plant Data Center, Baton Rouge, LA 70874-4490, USA.

Usher, M.B., Kruger, F.J., MacDonald, I.A.W., Loope, L.L, Brockie, RE. 1988. The ecology of biological invasions into nature reserves: an introduction. *Biol. Conserv.* 44, 1–8.

van der Heijden, M.G.A., Bardgett, R.D., van Straalen. 2008. The unseen majority: soil microbes as drivers of plant diversity and productivity in terrestrial ecosystems. *Ecology Letters* 11: 296–310.

Van der Heijden, MGA. et al. 1998. Mycorrhizal fungal diversity determines plant biodiversity, ecosystem variability and productivity. *Nature* 396: 69–72.

van der Putten, W.H., Klironomos, J.N. and Wardle, DA. 2007. Microbial ecology of biological invasions. *The ISME Journal* 1, 28–37.

Van der Putten, WH. 2002. How to be invasive. *Nature* 417: 32-33.

Van Devender, T.R., Felger, R.S., Bu´ rquez-Montijo, A., 1997. Exotic plants in the Sonoran Desert region, Arizona and Sonora. *Proceedings of the 1997 California Exotic Pest Plant Council Symposium.*

Van Wilgen, B.W., Cowling, R.M., Burgers, CJ. 1996. Valuation of ecosystem services: a case study from South African fynbos ecosystems. *BioScience* 46: 184–189.

van Wilgen, B.W., Reyersa, B., Le, Maitrea, D.C., Richardsonb, D.M., Schonegevel, LA. 2007 (Updated 2008). biome-scale assessment of the impact of invasive alien plants on ecosystem services in South Africa. *Journal of Environmental Management* 89(4):336-349.

VanKleunen, M. et al. 2010. Areinvadersdifferent? A conceptual framework ofcomparative approaches forassessing determinants of invasiveness. *Ecol. Lett.* 13: 947–958.

Vavra, M., Parks, C.G., Wisdom, MJ. 2007. Biodiversity, exotic plant species, and herbivory: The good, the bad, and the ungulate. *Forest Ecology and Management* 246: 66–72.

Vermeij, GJ. 1996. An agenda for invasion biology. *Biol. Conserv.* 78: 3–9.

Vilá, M. & Muñoz, I. 1999. Patterns and correlates of exotic and endemic plant taxa in the Balearic Islands. *Ecologia Mediterranea* 25: 153–161.

Vila`, M., & Weiner, J. 2004. Are invasive plant species better competitors than native plant species? evidence from pair-wise experiments. *Oikos* 105: 229-238.

Vilatersana, R., Brysting, A.K., Brochmann, C. 2007. Molecular evidence for hybrid origins of the invasive polyploids *Carthamus creticus* and *C. turkestanicus* (Cardueae, Asteraceae). *Molecular Phylogenetics and Evolution* 44: 610–621.

Villaseñor, J.L., Espinosa-Garcia, F.J. 2004. The alien flowering plants of Mexico. *Diversity and Distributions* 10: 113–123.

Vitousek, P.M., Lawrence, R.W., Whiteaker, L.D., Mueller-Dombois, D., Matson, PA. 1987. Biological Invasion by *Myrica faya* Alters Ecosystem Development in Hawaii. *Science* 238: 802-804.

Vitousek, P.M., Dantonio, C.M., Loope, L.L., Westbrooks, R. 1996. Biological invasions as global environmental change. *American Scientist* 84: 468–478.

Vitousek, P.M., Mooney, H.A., Lubchenco, J., Melillo, J.M. 1997. Human domination of earth's ecosystems. *Science* 277: 494–499.

Vitousek, P.M., Walker, L.R. 1989. Biological invasion by *Myrica faya* in Hawaii: Plant demography, nitrogen fixation, and ecosystem effects. *Ecological Monographs* 59: 247–265.

Vivanco, J.M., Bais, H.P., Stermitz, F.R., Thelen, G.C., Callaway, RM. 2004. Biogeographical variation in community response to root allelochemistry: novel weapons and exotic invasion. *Ecology Letters* 7: 285–292.

Volin, J.C., Lott, M.S., Muss, J.D., Owen, D. 2004. Predicting rapid invasion of the Florida Everglades by Old World Climbing Fern (*Lygodium microphyllum*). *Diversity Distrib.* 10: 439–446.

Von der, Lippe. M., Kowarik, I. 2007. Long-Distance Dispersal of Plants by Vehicles as a Driver of Plant Invasions. *Conservation Biology* 21 (4): 986–996.

W. R. Courtenay Jr., D. A. Hensley, J. N. Taylor, J. A. McCann. 1980. *The Zoogeography of North American Fishes*, C. H. Hocutt and E. O. Wiley, Eds. (Wiley, New York, 1986), pp. 675– 698

Walford, L., Wicklund, R. 1973. *FAO Fisheries Tech. Pap. 121* (1973); R. L. Welcomme, *FAO Fisheries*

Walther, G-R. 2003. Plants in a warmer world. *Perspectives in Plant Ecology, Evolution and Systematics* 6: 169–185.

Wang, R., Wang, YZ. 2006. Invasion dynamics and potential spread of the invasive alien plant species *Ageratina adenophora* (Asteraceae) in China. *Diversity Distrib.* 12: 397 –408.

Ward, N.L., Masters, GJ. 2007. Linking climate change and species invasion: an illustration using insect herbivores. *Global Change Biology* 13: 1605– 1615.

Wardle, D. 2001. Experimental evidence that plant diversity reduces invasibility - evidence of a biological mechanism or a consequence of sampling effect. *Oikos* 95: 161–170.

Wardle, D.A., Bardgett, R.D., Klironomos, J.N., Seta¨la, H., van der Putten, W.H., Wall, DH. 2004. Ecological Linkages Between Aboveground and Belowground Biota. *Science* 304 (11): 1629-1633.

Wardle, D.A., Nicholson, K.S., Ahmed, M., Rahman, A. 1994. Interference effects on the invasive plant Carduus nutans L. against the nitrogen-fixation ability of *Trifolium repens* L. *Plant Soil 163* (2): 287–297.

Wardle, D.A., Nicholson, K.S., Rahman, A. 1995. Ecological effects of the invasive weed species *Senecio jacobaea* L. (ragwort) in a New Zealand pasture. *Agriculture, Ecosystems and Environment* 56: 19–28.

Wardle, D.A., Nicholson, K.S., Rahman, A. 1995. Ecological effects of the invasive weed species *Senecio jacobaea* L. (ragwort) in a New Zealand pasture. *Agriculture Ecosystems and Environment* 56: 19-28.

Wardle, DA. 2001. Experimental demonstration that plant diversity reduces invasibility – evidence of a biological mechanism or a consequence of sampling effect? *Oikos* 95: 161–170.

Wardle, DA. 2002. *Communities and Ecosystems: Linking the Aboveground and Belowground Components*. Princeton University Press, Princeton, NJ.

Weber, E., Gut, D. 2004. Assessing the risk of potentially invasive plant species in central Europe. *Journal for Nature Conservation* 12: 171—179.

Wedin, D., Tilman, D. 1993. Competition among grasses along a nitrogen gradient: initial conditions and mechanisms of competition. *Ecological Monographs* 63: 199–229.

Wedin, D.A. & Tilman, D. 1996. Influence of nitrogen loading and species composition on the carbon balance of grasslands. *Science* 274: 1720–1723.

Weir, T.L., Park, S.W., Vivanco, J.M. 2004: Biochemical and physiological mechanisms mediated by allelochemicals. *Curr. Opin. Plant Biol.* 7: 472—479.

Weltzin, J.F., Muth, N.Z., Von Holle, B., Cole, PG. 2003. Genetic diversity and invasibility: a test using a model system with a novel experimental design. *Oikos* 103: 505–518.

Weste, G., Brown, K., Kennedy, J., Walshe, T., 2002. *Phytophthora cinnamomi* infestation — a 24-year study of vegetation change in forests and woodlands of the Grampians, Western Victoria. *Aust. J. Bot.* 50 (2): 247–274.

Western, D. 2001. Human-modified ecosystems and future evolution. *Proc. Nati. Acad Sci* 98 (10): 5458–5465.

Westman, WE. 1990. Park management of exotic plant species: problems and issues. *Conservation Biology* 4: 251–259.

Weston, L.A., Burke, B.A., Putnam, A.R. 1987: Isolation, characterization and activity of phytotoxic compounds from quackgrass [*Agropyron repens* (L.) Beauv.]. *J. Chem. Ecol.* 13: 403-421.

White, P.S. Pickett, S.T.A.. 1985. Natural disturbance and patch dynamics: An introduction. In: S.T.A. Pickett and P.S. White, (Editors), *The Ecology of Natural Disturbance and Patch Dynamics*. Academic Press, Orlando, FL, pp. 3-13.

White, T.A., Campbell, B.D., Kemp PD. 1997. Invasion of temperate grassland by a subtropical annual grass across an experimental matrix of water stress and disturbance. *Journal of Vegetation Science* 8: 847–854.

Whittaker, R.H., Feeney, PP. 1971. Allelochemics: chemical interactions between species. *Science* 17: 757–770.

Wiesemeier, T., Hay, M., Pohnert, G. 2007. The potential role of wound-activated volatile release in the chemical defence of the brown alga *Dictyota dichotoma*: Blend recognition by marine herbivores. *Aquat. Sci.* 69:403 – 412.

Wilcove, D.S., Rothstein, D., Dubow, J., Phillips, A., Losos, E. 1998. Quantifying threats to imperiled species in the United States. *Bioscience* 48: 607–615.

Williams, J.A., West, CJ. 2000. Environmental weeds in Australia and New Zealand: issues and approaches to management. *Austral Ecology* 25: 425–444.

Williams, K., Hobbs, R.J., Hamburg. SP. 1987. Invasion of an annual grassland in Northern California by *Baccharis pilularis* ssp. *consanguinea. Oecologia,* 72: 461-465.

Williamson, GB. 1990. Allelopathy, Koch's postulates, and the neck riddle. In: Grace JB, Tilman D, eds. *Perspectives on plant compostion.* New York: Academic Press, 143–162.

Williamson, M. 1996. *Biological Invasions.* Clapman and Hall, London, UK, 244 pp.

Williamson, M. 1999. Invasions. *Ecography* 22: 5-12.

Williamson, M., Fitter, A. 1996a The varying success of invaders. *Ecology* 77: 1661–1666

Williamson, M.H., Fitter, A. 1996.b The characters of successful invaders. *Biological Conservation* 78, 163–170.

Willis, A.J., Memmott, J., Forrester, RI. 2000. Is there evidence for the post invasion evolution of increased size among invasive plant speies? *Ecology Letters* 3: 275-283.

With, K.A. 2001. The landcape ecology of invasive spread. *Conservation Biology* 16(5): 1193-1203.

Witkowski, E.T.F., Garner, R.D. 2008. Seed production, seed bank dynamics, resprouting and long-term response to clearing of the alien invasive *Solanum mauritianum* in a temperate to subtropical riparian ecosystem. *South African Journal of Botany* 74: 476–484.

Witta, A.B.R., McConnachiea, A.J., Stals, R. 2004. *Alcidodes sedi* (Col.: Curculionidae), a natural enemy of *Bryophyllum delagoense* (Crassulaceae) in South Africa and a possible candidate agent for the biological control of this weed in Australia. *Biological Control* 31: 380–387.

Wolfenbarger, L.L., Phifer, P.R. 2000. The Ecological Risks and Benefits of Genetically Engineered Plants. *Science* 290: 2088-2093.

Wu, S.H., Hsieh, C., Chaw, S., Rejmánek, M. 2004. Plant invasions in Taiwan: Insights from the flora of casual and naturalized alien species. *Diversity Distrib.* 10: 349–362.

Xu, C., Gertner, G.Z., Scheller, R.M. 2007. Potential effects of interaction between CO_2 and temperature on forest landscape response to global warming. *Global Change Biology* 13: 1469–1483.

Yachi, S., Loreau, M. 1999. *Proc. Natl. Acad. Sci.* 96, 1463.

Yeates, G.W., Boag, B. 2001. Background for nematode ecology in the 21st century. In: Chen,Z. X., Chen, S. Y., Dickson, D. W. (eds) *Nematology, Advances and Perspectives*. Tsinghua University Press, China.

Yeates, G.W., Bongers, T. 1999. Nematode diversity in agroecosystems. Agriculture, *Ecosystems and Environment* 74, 113–135.

Yeates, G.W., Williams, P.A. 2001. Influence of three invasive weeds and site factors on soil microfauna in New Zealand. *Pedobiologia* 45: 367–383.

Yong Sunga, C., Li, M., Rogers, G.O., Volder, A., Wang, Z. 2011. Investigating alien plant invasion in urban riparian forests in a hot and semi-arid region. *Landscape and Urban Planning* 100: 278–286

Young et al. 1998. The interaction of soil biota and soil structure under global change. *Global Change Biology* 4: 703-712.

Zalba, S.M., Sonaglioni, M.I., Compagnoni, C.A., Belenguer, C.J. 2000. Using a habitat model to assess the risk of invasion by an exotic plant. *Biological Conservation* 93: 203-208.

Zancola, B.J., Wild, C. and Jena, C.M. 2000. Inhibition of *Ageratina riparia* (Asteraceae) by native Australian flora and fauna. *Austral Ecology* 25: 563–569.

Zangerl, A.R., Berenbaum, M.R. 2005. Increase in toxicity of an invasive weed after reassociation with its coevolved herbivore. *Proc. Nati. Acad Sci* 102(43): 15529–15532.

Zangerl, A.R., Stanley, M.C., Berenbaum, M.R. 2008. Selection for chemical trait remixing in an invasive weed after reassociation with a coevolved specialist. *Proc. Nati. Acad Sci* 105 (12): 4547–4552.

Zavaleta, E.S., Hulvey, K.B. 2004. Realistic Species Losses Disproportionately Reduce Grassland Resistance to Biological Invaders. *Science* 306: 1175-1177.

Zavaleta, E. S. 2000.in *Invasive Species in a Changing World* (eds Hobbs, R. J. & Mooney, H. A.) (Island Press Washington DC), pp.457.

Zavaleta, E.S., Hobbs, R.J. and Mooney, H.A. 2001. Viewing invasive species removal in a whole-ecosystem context. *Trends in Ecol. Evol.*16(8), 454-459.

Zavaleta, E.S., Shaw, M.R., Chiariello, N.R., Mooney, H.A. and Field, C.B. 2003. Additive effects of simulated climate changes, elevated CO_2, and nitrogen deposition on grassland diversity. *Proc. Nati. Acad Sci* 100 (3): 7650–7654.

Zhang, L.Y., Ye, W.H., Cao, H.L., Feng, H.L. 2004. *Mikania micrantha* H. B. K. in China – an overview. *Weed Research* 44: 42–49.

Zibrowius, H. 1991. *Ongoing modi¢cation of the Mediterranean marine fauna and flora by the establishment of exotic species* 1991 Me´ soge´ e 51: 83.

Index

T

U